PROTECTING WASTEWATER PROFESSIONALS FROM COVID-19 AND OTHER BIOLOGICAL HAZARDS

T0344686

2020

Water Environment Federation
601 Wythe Street
Alexandria, VA 22314-1994 USA
http://www.wef.org

About WEF

The Water Environment Federation (WEF) is a not-for-profit technical and educational organization of 33,000 individual members and 75 affiliated Member Associations representing water quality professionals around the world. Since 1928, WEF and its members have protected public health and the environment. As a global water sector leader, our mission is to connect water professionals; enrich the expertise of water professionals; increase the awareness of the impact and value of water; and provide a platform for water sector innovation. To learn more, visit www.wef.org.

Contents

List of Figures

List of Tables

Executive Summary

The unprecedented global pandemic of SARS-CoV-2 virus causing the COVID-19 disease prompted the Water Environment Federation (WEF) to conduct a critical review of pathways of potential exposure to this virus associated with the collection and treatment of wastewater.

In April 2020, WEF convened a panel of academics and practitioners expert in the science and practice of wastewater collection and treatment to conduct this critical review, which included an assessment of guidance found in WEF publications and other publicly available sources. The goal of the blue-ribbon panel's review was to

1. Ensure that WEF provides the most current, evidence-based information on protecting worker health and safety;

2. Provide appropriate input to the technical medical and public health community, including governmental agencies such as the Centers for Disease Control and Prevention (CDC), the Occupational Safety and Health Administration (OSHA), the World Health Organization (WHO), and parallel authorities in other countries; and

3. Ensure that worker health and safety guidance provided is consistent and accurately reflects the knowledge and insights of the wastewater sector.

1

The blue-ribbon panel was also charged with producing three deliverables:

1. Updates as necessary and appropriate to *Safety, Health, and Security in Wastewater Systems*, or MOP 1 (WEF, 2012), to reflect current knowledge on best practices for protecting health and safety workers from biological hazard exposures associated with wastewater;
2. A review of worker health and safety best practice guidelines from agencies and authorities external to WEF and recommendations to supplement said guidelines, specifically with respect to exposure to biological hazards associated with wastewater; and
3. A review of current, relevant peer-reviewed literature to identify and report on gaps in the knowledge base specific to exposure to biological hazards associated with wastewater for which research is needed to further inform best practices for the wastewater sector.

The blue-ribbon panel's final peer-reviewed deliverables are included in their entirety in this document. This executive summary provides highlights specific to the SARS-CoV-2 virus.

The following is the registry of blue-ribbon panel participants and their affiliations:

Name	Role	Affiliation
Art Umble	Chair	Stantec Consulting, Inc.
Allegra Da Silva	Vice Chair	Brown and Caldwell
Tim Page-Bottorff	2nd Vice Chair	Safestart
David Gill	Panelist	DC Water
Mark LeChevallier	Panelist (Workgroup Lead)	Dr. Water Consulting
Naoko Munakata	Panelist	Los Angeles County Sanitation District
Kyle Bibby	Panelist	University of Notre Dame
Mark Sobsey	Panelist (Workgroup Lead)	University of North Carolina
Chuck Gerba	Panelist	University of Arizona
Chuck Haas	Panelist (Workgroup Lead)	Drexel University
Amy Kirby	Panelist	Centers for Disease Control
John Bannen	Panelist	Inframark
Leonard Casson	Panelist	University of Pittsburgh

Chip Blatchley	Panelist	Purdue University
Kartik Chandran	Panelist	Columbia University
Sharon Nappier	Panelist	U.S. Environmental Protection Agency
Claudio Ternieden	Sr. Director, Government Affairs & Strategic Partnerships	Water Environment Federation
Barry Liner	Chief Technical Officer	Water Environment Federation
Elizabeth Conway	Technical Project Manager	Water Environment Federation
Lorna Ernst	Senior Director, Publishing	Water Environment Federation
Tim Williams	Deputy Executive Director	Water Environment Federation
Patrick Dube	Biosolids Practice Lead	Water Environment Federation

1.0 REVISIONS AND UPDATES TO *SAFETY, HEALTH, AND SECURITY IN WASTEWATER SYSTEMS*

From a public health and safety perspective, an important consideration of viruses is that they reproduce only in living cells and cannot reproduce without a host cell. Therefore, viruses will not reproduce in wastewater. A primary source of viruses that are infectious to humans is human waste discharged to the collections system. There are more than 100 different types of viruses found in human waste. Non-enveloped viruses, such as enteric viruses, are more persistent in wastewater, and therefore wastewater personnel have a higher potential for exposure to these viruses because of their daily contact with wastewater. Conversely, enveloped viruses (e.g., coronavirus), are more sensitive to wastewater treatments, and thus are less of a concern to people in contact with wastewater.

SARS-CoV-2 is an enveloped virus and less stable in the environment. The main routes of transmission are respiratory droplets and direct contact, with no reports of fecal-oral transmission to date, and the risk of transmission of SARS-CoV-2 from feces appears to be low. Standard treatment and disinfectant processes at wastewater treatment plants are expected to be effective at inactivating SARS-CoV-2. Best practices for protecting the health of workers exposed to wastewater should be followed regardless, such as:

engineering and administrative controls; safe work practices; and personal protective equipment (PPE) normally required for tasks when handling untreated wastewater. At locations where wastewater or sludge is sprayed, the possibility of inhaling infectious agents will increase. In instances where exposure to aerosols is anticipated, use of N95 respirators or surgical masks and goggles may help to minimize contact. No additional COVID-19–specific protections are recommended. A PPE summary is given in Table E.1. In addition, it is also important to practice proper personal hygiene to reduce the possibility of contamination by wastewater.

Because of their direct high exposure to raw wastewater, collection system workers (including combined sewer overflow and sanitary sewer overflow operators) have greater risks of infection from pathogenic agents present in wastewater, as compared to treatment facility personnel, as found in studies. Furthermore, wastewater personnel engaged in biosolids handling, laboratory analytics, septic haulers, landfill leachate handlers, industrial pretreatment personnel, and persons handling ambient water quality sampling all have potential exposure. But these risks resulting directly from SARS-CoV-2, specifically, are considered low, because this virus in its infectious form, has not, as of yet, been detected in wastewater. In all cases, application of appropriate PPE and hygiene practices can reduce these risks.

2.0 KEY SAFETY INFORMATION AND CONSIDERATIONS FOR COVID-19 AND OTHER BIOHAZARDS

Based on the available data on the SARS-CoV-2 virus and knowledge about similar viruses, experts agree that the occupational risk of infection to wastewater workers from the SARS-CoV-2 virus is low, and not greater than

TABLE E.1 Personal protective equipment summary.

Route of exposure	Recommended PPE
Contact transfer	Gloves, boots, and uniform/coverall
Splash	Protection of the eyes/face through safety glasses, face shield, or goggles
Whole-body contact	Tyvek suits or coveralls
Abrasion, cut, or puncture	Durable gloves designed for protection from cuts or punctures
Respiratory from sprays, mists, or dust	N95 respirator, surgical or dust mask

those from other pathogens typically present in wastewater. The following are key points.

- The SARS-CoV-2 virus is mainly transmitted through respiratory droplets produced when an infected person coughs, sneezes, or talks, or by direct contact with an infected person, with possible further transmission by contact with contaminated surfaces (e.g., from deposited respiratory droplets).

- Transmission via wastewater aerosols, fecal wastes, wastewater, sludge, or biosolids is unlikely. Risks of transmission are site- and job-specific: the type and level of hazards vary by task and by the conditions, equipment, and configuration at each utility. Taking the precautions that protect workers against typical wastewater pathogens should also adequately protect workers against the SARS-CoV-2 virus.

- Infectious SARS-CoV-2 virus has not been detected in wastewater, sludge, and biosolids thus far; although the presence of infectious SARS-CoV-2 virus in these environments cannot be ruled out, it appears unlikely to be present at concentrations that would cause a measurable health risk.

- The risk of contracting SARS-CoV-2 virus from exposure to wastewater, sludge, and biosolids is considered low, due to its unlikely presence and the expected dilution and die-off of the virus in those matrices. The risk from aerosols in wastewater and biosolids systems is similarly low for these reasons.

- If the SARS-CoV-2 virus is found to be present in a wastewater matrix, treatment processes for wastewater and biosolids are expected to significantly remove and/or inactivate this virus.

- The SARS-CoV-2 virus is not expected to pose any new or increased risks for infection. Workers in the wastewater environment should protect themselves by consistently and properly applying appropriate safety measures and best practices.

- Default personal protective equipment (PPE) includes waterproof gloves, waterproof boots, a uniform or waterproof coveralls, a covering for the mouth and nose (medical grade mask or respirator), and eye protection (e.g., safety glasses, goggles, or a face shield).

- Medical or dust masks can protect workers from splashes, sprays, and particulate materials (e.g., dust), and reduce the chances that sick workers will spread their illnesses.

- A Job Safety (Hazard) Analysis (JSA or JHA) should be used to determine conditions where a respirator (such as an N95 filtering facepiece) might be appropriate for protection from inhalation of aerosols.

- Because the specific biological hazards vary by job activity and wastewater system, it is necessary for facility managers and staff to conduct a Job Safety (or Hazard) Analysis (JSA/JHA) (OSHA, 2002). This analysis identifies each task in a job, defines the potential hazards, and outlines critical safety practices.

3.0 RESEARCH NEEDS RELATED TO SARS-COV-2

3.1 What We Know That We Do Not Know About SARS-CoV-2

- Little is currently known of the partitioning of viruses between solid and liquid phases in wastewater. This is particularly true for the SARS-CoV-2 virus. Knowledge in this area must be inferred from other viruses.
- If, in the wastewater environment, the majority of SARS-COV-2 partitions to the solids, then methods optimization should focus on most efficient solids separation and recovery from the solid concentrate.
- Viral association with particles may serve to protect the viruses from disinfectants either by shielding, in the case of ultraviolet (UV) disinfection, or by hindering chemical penetration.
- When associated with particles during treatment, or in receiving waters, the transport of virus will be governed by particle behavior. Therefore, hydrodynamic models of transport, as well as treatment process models, must consider such associations.
- It is likely that association of viruses with particles is non-specific relative to the host-receptor-specific binding. Applicability to SARS-CoV-2 should be confirmed.
- Methodologies exist to remove polymerase chain reaction (PCR) detections of nucleic acids from damaged virions, potentially providing a pathway to connect molecular and infectious viral quantifications. Though these methodologies have recently been reviewed, there is no awareness of prior work demonstrating their suitability for environmental detections of enveloped viruses, such as SARS-CoV-2.
- Critical work is necessary to determine if viable SARS-CoV-2 is present in wastewater and at what concentrations. Additional approaches to remove PCR detections of damaged SARS-CoV-2 virions may also help to unify these measurements.

- Minimum data reporting requirements should be developed. While measurements of raw wastewater and sludge may be useful for population surveillance, there may be limited value to analyzing final effluents, and this needs further exploration.
- No data are yet available on SARS-CoV-2 persistence or disinfection kinetics in the water environment. However, there has been prior work on survival of SARS-CoV-1 in aqueous systems. It is critical that future experiments investigating SARS-CoV-2 persistence and disinfection assess viable virus.
- While data on SARS-CoV-2 persistence and disinfection in water matrices are currently limited, we may be able to gain insights into likely viral fate using surface persistence data.
- Though it is believed that current wastewater disinfection processes are likely to be effective against coronaviruses, quantitative data for comparison to the more commonly studied enteric viruses are lacking and are needed to have the assurances that treatment is sufficient to control the risk from coronaviruses.
- No data are yet available on SARS-CoV-2 to define UVC (UV light with wavelengths between 200 and 280 nm) dose-response behavior (i.e., disinfection kinetics) of SARS-CoV-2 in aqueous suspension or on surfaces.
- No data are available to define the action spectrum (i.e., wavelength-dependence) of SARS-CoV-2 for UVC radiation.
- No data appear to be available for inactivation in water by chlorine dioxide or ozone.
- Information specifically on removal of enveloped viruses, such as SARS-CoV-2, in various treatment processes and trains used for water reuse is not available.

3.2 Key Recommendations for Immediate Research Related to SARS-CoV-2

1. Develop, fund, and conduct a prospective epidemiological study of infectious disease incidence among wastewater and collection system workers. The study should at a minimum
   - establish a baseline level of PPE use among wastewater and collection system workers; and
   - examine observed health outcomes with respect to reported PPE use to test the efficacy of PPE as currently practiced in reducing infectious disease risks.

2. Develop, fund, and conduct a study to characterize respiratory exposure for typical tasks performed by workers in wastewater collection and treatment operations. The presence and magnitude of infectious agents in tasks performed by wastewater workers is not well understood, even though emerging research indicates that aerosolization of wastewater may expose them.

1

Revisions and Updates to *Safety, Health, and Security in Wastewater Systems*

Common routes of exposure, risks, and prevention guidelines associated with common water resource recovery facility hazards are shown below in Table 1.1.

1.0 TYPES OF HAZARDS

1.1 Bacteria

Pathogenic bacteria are microscopic and are extremely common in wastewater. Because bacteria can reproduce outside the body, microorganisms

TABLE 1.1 COMMON WATER RESOURCE RECOVERY FACILITY HAZARDS.

Adapted from: *Safety, Health, and Security in Wastewater Systems, Manual of Practice No. 1*, Water Environment Federation, 2012 Chapter 4: Safety and Health in Wastewater Treatment Plant Operation

Common exposures	Typical exposure sources/activities	Risks	Prevention guidelines	Other resources
Asbestos, Asbestos–cement pipe (ACP), Thermal insulation systems	Renovation, demolition, or other activities that damage or disturb asbestos-containing products such as insulation, fireproofing, acoustical materials, floor tiles, and asbestos–cement (transite) pipe.	When disturbed, microscopic asbestos fibers can become airborne and can be inhaled into the lungs, where they can cause serious lung diseases.	Maintain asbestos-containing materials (ACMs) in good condition through a well-run building maintenance program; undisturbed asbestos materials do not pose a health risk. When removing ACP, conduct an initial exposure assessment to demonstrate that employees' exposures are below the OSHA 8-hour time-weighted average PEL of <0.1 fibers/cubic centimeter and follow OSHA abatement guidelines for removal. Contact an asbestos professional for consultation and removal if exposures will exceed the OSHA PEL.	See 29 CFR 1926.1101 for regulatory details.
Biological hazards (not to include bloodborne pathogens)	Exposure to untreated or partially treated wastewater, sludge or biosolids through direct contact, splashes, cuts, abrasions, or aerosols.	Acute infections from viruses, bacteria, protozoa, fungi, or helminths.	A Job Safety (or Hazard) Analysis needs to be completed for each activity to determine the probability, severity, and potential consequence of each hazard and the appropriate hazard control measures. PPE can include gloves, boots, uniform/coveralls/ waterproof suits, eye protection, face shield or goggles, N95 respirator, surgical or dust mask.	Refer to Chapter 8 for additional information.*
Chemicals (liquid), Sodium hypochlorite, sodium bisulfite, sodium hydroxide, ferric chloride, polymers, anhydrous ammonia, hydrogen peroxide, and methanol	Unloading, connecting, and disconnecting activities; transferring bulk chemicals to secondary containers; and emergency response to accidental spills.	Acute respiratory hazards from inhalation of toxic offgases or vapors; potential acute or chronic injuries/illness to unprotected eyes and skin; slip hazards from wet, slippery surfaces; and potential thermo–dynamic reactions from mixing incompatible materials	Employees must be trained on proper chemical handling procedures and wear the proper level of PPE (refer to the chemical's safety data sheet [SDS] for PPE recommendations); employees who will respond to a spill must be trained in emergency response procedures and be equipped with the proper level of PPE; clean up all spills immediately, no matter how small; and properly label bulk chemicals including tanks, associated pipes, pumps, and valves.	Refer to 4.2.6, 4.3.3, 4.4, 4.4.3 &, and Chapter 10 for additional information.*
Chlorine (gas)	Loading/unloading, connect and disconnect activities, and emergency response to accidental releases.	Acute respiratory hazards from inhalation of chlorine gas. At 40 to 60 ppm, chlorine gas causes toxic pneumonitis and pulmonary edema; at 430 ppm, an exposure is lethal; and, in 30 minutes and at 1000 ppm, it is fatal within minutes. Local chemical or thermal (frostbite) burns to the skin from contact with liquid chlorine. Gaseous chlorine in contact with the skin can dissolve in body moisture to form hypochlorous and hydrochloric acids. Eye irritation from low concentration of chlorine gas to serious thermal and/or chemical burns from liquid chlorine.	Employees must be trained on chlorine handling procedures and must wear the proper level of PPE when working with chlorine. During loading/unloading procedures, either a full-face air-purifying respirator or self-contained breathing apparatus (SCBA) should be worn depending on the expected exposure concentrations. Emergency response to a release will require a higher level of protection. In addition to gloves (thermal protection) and a SCBA, the Chlorine Institute recommends the following: level B PPE when liquid chlorine is not involved and enhanced Level B PPE when liquid chlorine is involved.	Refer to 4.2.6.1 and Chapter 10 for additional information.*
Compressed gases	Storing and using compressed gas and air cylinders.	Projectile hazards from sudden release of pressures; fire and explosions from flammable or reactive gases; oxygen-deficient atmospheres from leaks of inert gases; and poisonous atmospheres from leaks of toxic gases.	Properly secure cylinders in a well-ventilated area away from heat, flames, and the sun; when not in use, store upright and leave cap on and valve closed; when stored inside, store compressed gases at least 6.1 m (20 ft) from highly combustible materials; separate oxygen cylinders from fuel gas cylinders at least 1 minimum of 6.1 m (20 ft) or by a non-combustible barrier at least 1.5 m (5 ft) high; and SCBA breathing air cylinders must be tested and maintained as prescribed in the Shipping Container Specification Regulations of the U.S. Department of Transportation (49 CFR part 173 and part 178).	See 29 CFR, 1910 Subpart H for compressed gas regulatory details. See 29 CFR 1910.134 for breathing air regulatory details.
Confined spaces	Entering (i.e., breaking the plane) into a confined space.	Toxic atmospheric conditions from hydrogen sulfide, carbon monoxide, and other contaminants; oxygen-deficient conditions; and flammable conditions.	No one may enter a confined space unless properly trained as an entrant, follow confined space entry procedures including ventilation, gas monitoring, lockout/blockout procedures, and using the proper level of PPE and rescue equipment.	Refer to Chapter 5.8 and Chapter 6.5 for additional information.* See 29 CFR 1910.146 for regulatory details.
Cranes and hoists	Using a mobile crane or a hoist to lift and/or move materials.	Cranes: Fatalities and serious injuries from a crane tip-over; electrocution from accidental contact with overhead lines; and injuries and/or damages from accidental contact with people, equipment, or buildings. Hoists: Pinch/nip points and injuries and/or damages from dropping the load.	Only employees who are authorized by the employer and trained in safe operations of cranes or hoisting equipment may operate such equipment. In certain states, mobile crane operators must have a valid certificate of competency (certificate) by an accredited certifying entity to operate mobile cranes meeting certain load capacity and other criteria.	See 29 CFR 1926.552 and 1910.179 for regulatory details.
Elevated work	Working from unprotected elevated locations; elevating work platforms and aerial lifts, ladders, and scaffolds; and working near unprotected wall or floor openings.	Fatalities, broken bones, spinal cord injuries, concussions, and other serious injuries as a result of falling off of an elevated work area or into a pit or other opening. According to the Bureau of Labor Statistics, in 2007, 15% of all fatalities were from falls.	Ensure all elevated work areas have standard guardrails if they expose employees to a fall of 1.2 m (4 ft) or more; only use ladders that are properly rated for the load and type of work being performed; inspect ladders before use for damages or wear; ensure scaffolding has been erected and dismantled by a qualified person, do no overload scaffolds and maintain them in a safe condition; employees who operate or drive an elevating work platform or aerial lift must be trained; use a personal fall protection system if standard guardrails are not provided; keep all floor and wall openings covered or protected by a standard guardrail or safe effective barrier; never leave an open manhole unattended or other pit unattended; and ensure skylight openings are guarded by a standard skylight screen or have a fixed standard railing on all exposed sides.	Refer to Chapter 5.1 for additional information.* See 29 CFR 1910 Subpart D; 29 CFR 1910 Subpart F; 29 CFR 1926 Subparts L–N for regulatory details.

Common exposures	Typical exposure sources/activities	Risks	Prevention guidelines	Other resources
Excavation and trenching	Cutting, trenching, or making depressions into the surface and removing earthen materials in order to lay or repair pipes, lines, cables, or other underground installations.	Suffocation from cave-ins; toxic or oxygen-deficient atmospheric conditions; explosions and fires from cutting live fuel lines; and electrocutions from cutting live electrical lines.	Daily inspections of excavations, the adjacent areas, and protective systems by a competent person; installing protective systems to protect employee from falling materials and cave-ins; marking the location of all known underground utility installations; providing safe access and egress, providing protection from vehicular traffic, falling loads and water accumulation; and testing the atmosphere if employees enter an excavation greater than 1.2 m (4 ft) if oxygen deficiency or hazardous atmospheres exist (or could reasonably be expected to exist.)	Refer to Chapter 6.4 for additional information.* See 29 CFR 1926 Subpart P for regulatory details.
Fires and explosions	Fueling vehicles and equipment, storing and using flammable materials, and working on, or near, flammable fuel sources such as methane or natural gas supplies.	Fires and explosions.	Turn off engine before refueling any internal combustion engine with a flammable liquid; keep open lights, flames, or sparking/arcing equipment away from flammable areas, including fuel tanks; ensure flammable containers are plainly marked as a flammable substance; keep flammable liquids in covered containers when not in use; store flammable substances in an approved flammable-storage cabinet; provide portable fire extinguishment and control equipment in quantities and types needed for the types of hazards presented; and use grounding and bonding techniques for all Class I liquids when dispensing into containers unless the nozzle and container are electrically interconnected.	See 29 CFR 1910.1.6 for regulatory details.
Flying particles	Performing activities such as grinding, chipping, landscaping and using a jackhammer or other equipment that cause flying particles to be emitted.	Eye and skin injuries.	When feasible, use shields, screen, guards or enclosures and wear eye, face, and body protection.	Refer to Chapter 10 for additional information.* See 29 CFR 1910.133.
Hazardous energy	Cleaning, repairing, servicing, setting-up, or adjusting prime movers or electrical systems.	Electrocutions, amputations, or other serious injuries from unexpected energization or startup of machines or equipment or from the release of stored or secondary energy (e.g., capacitors, springs, batteries, and uninterrupted power supplies).	Only trained employees may work on equipment, machines, or electrical systems under lockout/tagout conditions and follow lockout/blockout/tagout procedures when cleaning, repairing, servicing, setting up, or adjusting prime movers or electrical systems.	Refer to Chapter 5.2 for additional information.* See 29 CFR 1910.147 for regulatory details.
Heat and cold	Working outdoors in extreme weather conditions for prolonged periods of time.	Heat illnesses including heat stroke, heat exhaustion, heat cramps, fainting, and heat rash and hypothermia and frostbite.	Provide adequate shade and water for employees who need to work in extreme hot weather conditions and train them on heat illness prevention and provide short exposures, warm clothing, and hot beverages to employees who need to work in extreme cold weather conditions and train them on cold weather illness/injury prevention.	Refer to http://www.osha.gov/index.html for heat stress publications.
Hot surfaces	Working near pipes or other exposed surfaces that are within 2.1 m (7 ft) and have an external surface temperature of 60 °C (140 °F) or higher.	Burns.	Guard against contact or cover with thermal insulating material.	California Code of Regulations, Title 8, §3308.
Hot work	Welding, cutting, brazing, or performing any work activity that produces heat, flames, or sparks.	Fires and explosions.	Perform welding in an approved hot work area if possible, evaluate the surrounding work area for flammable or combustible products and implement safeguards to protect against accidental fires or explosions; do not perform hot work activities in oxygen-enriched conditions; test any area for oxygen if there is any opportunity for enriched oxygen atmospheric conditions (e.g., pure oxygen treatment systems); and complete a hot work permit when performing hot work inside a confined space.	Refer to Chapter 5.7 for additional information.* See 29 CFR 1910.146 for regulatory details when performing hot work in confined spaces.
Industrial trucks (forklifts)	Driving or operating an industrial truck (forklift).	Fatalities or serious injuries from tipping the forklift over; damages to equipment and structures; and carbon monoxide poisoning.	Employees may not drive or operate a forklift unless they have been trained and have shown competency in their ability to drive or operate the forklift; when operating internal combustion engine forklifts inside enclosed or semi-enclosed buildings or structures, ensure that natural or mechanical ventilation systems keeps toxic fumes and gases below the OSHA permissible limits; if unsure, wear a personal gas monitor while operating a forklift inside an enclosed work area.	Refer to Chapter 7 for additional information.* See 29 CFR 1910.178 for regulatory details.
Lead	Demolition or renovation activities that involve lead-based paints.	Lead poisoning.	If lead is present, perform an exposure assessment to determine if any employee may be exposed to lead at concentrations greater than 50 ug/m³ of air averaged over an 8-hour period. Wear the appropriate respiratory protection and other PPE to prevent exposure to lead.	Refer to Chapter 10 for additional information on respiratory protection.* See 29 CFR 1926.62 and 29 CFR 1910.1025 for regulatory details.
Line breaking	Opening closed-pressurized or gravity-fed pipes or systems containing hot, poisonous, corrosive, flammable, or other hazardous substances.	Burns and other injuries from exposures to chemicals, steam, and toxic or other hazardous products.	Identify what the pipe or systems contain before breaking any lines or pressurized systems; purge with water or other compatible substances to reduce the concentrations inside the pipe or system; and relieve the internal pressure to prevent the sudden release of pressure or spraying liquid.	See 29 CFR 1910.119, 29 CFR 1910.146 for regulatory details. California Code of Regulations, Title 8, §3329.
Machine guarding	Operating or working near prime movers, machines, and machine parts that grind, shear, punch, squeeze, cut, roll, mix, rotate, or similar actions.	Amputations or other serious injuries to the hands, fingers, arms, or other body parts.	Provide machine guarding if employees come within the danger zone.	Refer to Chapter 5.3 for additional information.* See 29 CFR 1910 Subpart 0.
Muscular skeletal disorders (ergonomics)	Manual material handling (MMH) work such as lifting, moving, pulling, pushing and using hand-held and stationary vibrating tools.	Strains and sprains to the back, shoulders, and upper limbs and vibration syndrome, also know as *white finger* and as *Raynaud's phenomenon of occupational origin.*	Manual material handling: Improve the fit between the demands of the work tasks and the capabilities of the workforce through engineering improvements (i.e., modifying or redesigning equipment, processes, or workstations), administrative improvements (i.e., alternating heavy tasks with light tasks and rotating job activities), and modifying work practices so that workers perform work within their power zone (i.e., above the knees, below the shoulders, and close to the body). Use alternatives to handling materials manually such as special tools and powered and nonpowered equipment. Vibration: Reduce the acceleration speed of the tool or equipment, keep tools and equipment well maintained, use energy-dampening tools to reduce transmission from the tool to the hand, and modify the process to reduce or eliminate the need for vibrating tools.	For additional information on manual material handling, go to http://www.cdc.gov/az/ DHHS (NIOSH) Publication No. 2007-131 (April 2007) and DHHS (NIOSH) Publication No. 89-106.

Common exposures	Typical exposure sources/activities	Risks	Prevention guidelines	Other resources
Noise exposures	Working in high-noise areas such as backup generator rooms, blower rooms, and compressor rooms and performing high-noise activities such as using landscaping tools and jackhammers and operating heavy equipment, vactor trucks, or portable pumps.	Loss of hearing.	Perform a complete noise survey on all work areas and job tasks that are suspect of noises that are higher than OSHA allowable limits, post areas that require hearing protection to be worn, wear hearing protection when required, and participate in a hearing conservation program if exposed to the OSHA action level of an 8-hour time-weighted average of 85 dB on the A scale (slow response) or, equivalently, a dose of 50%.	See 29 CFR 1910.95 for regulatory details.
Powder-actuated tools	Driving fasteners into concrete or steel walls using powder-actuated tools (tools that use powder cartridges).	Fatality or other serious eye and body injuries from nails, fasteners, or studs.	Only trained and qualified persons may operate powder-actuated tools. Keep the tool in a locked container when not in use. Post warning signs when in use. Only use fasteners and power loads recommended by the tool manufacturer.	Refer to Chapter 5.5 for additional information.* See CFR 1926.302 for regulatory details.
Power and hand tools	Operating electric, hydraulic, pneumatic, fuel-powered tools.	Electrical shock, effect from attachments accidentally being expelled; injuries from ricochets; injuries from high-pressure, high-velocity releases; and flammable or oxygen deficient conditions when using fuel-powered tools in enclosed work areas.	Only electric power-operated tools that are double-insulated or grounded should be used. Never remove or deactivate safety clips, retainers, and other safety devices on pneumatic impact (percussion) tools. Shut off fuel-powered tools when refueling or servicing. Wear a personal gas monitor when using fuel-powered tools in enclosed, or semi-enclosed, work areas. Do not exceed the manufacturer's safe operating pressures for hoses, valves, pipes, filters, and other fittings when using hydraulic tools.	Refer to Chapter 5.5 for additional information.* See CFR 1926.302 for regulatory details.
Radiological hazards	Natural and manmade radionuclides that are discharged into sanitary sewers; radiological dispersion event (RDE) leading to significant quantities of radioactive material into the combined sanitary and storm sewer system. An RDE could come from a radiological dispersion device such as a "dirty bomb" contaminating the community at large or from the deliberate and malicious introduction or dispersion of radioactive material into the waterways and water supply systems.	Studies have shown that natural and manmade radionuclides in wastewater sludge and ash do not indicate a widespread problem and that exposure doses are generally well below levels that would require radiation protection actions. Direct irradiation of wastewater workers is possible from a radiation dispersal event. Of particular concern is the malicious dispersal of radiation materials into the wastewater system through a radiological dispersion device.	Being prepared for an RDE into the sewer system will require extensive studies and detailed plans. Among other things, procedures for detecting radioactive exposures, whether to treat or bypass the waste, how to segregate, store, and dispose of radioactive wastes, plus the duty (if any) of wastewater workers to protect the public by treating contaminated wastes will have to be considered.	Strom, D. J. (2005) *Radiological Risk Assessment for King County Wastewater Division*, PNNL 15163 Vol 1. http://www.pnl.gov/main/publications/external/
Removing covers	Opening/removing a variety of covers such as manhole covers, tank covers, blind flanges, hatches, or grates.	Soft tissue injuries to back, shoulders, and arms. • Fall hazards from open pits • Toxic atmospheric conditions and/or flammable conditions from concentrated gases under the covers • Pinch points • Hazardous energy sources when opening pressurized systems • Slip/trip hazards from loose or uneven covers and gratings	Use lifting devices designed for mechanical advantage to prevent soft tissue injuries when removing heavy covers. When covers are removed, provide an attendant or a physical barrier (e.g., portable guardrail) to prevent falls into floor openings or pits. Perform an atmospheric test before opening a cover if there is a potential for buildup of flammable or toxic gases underneath the lid or inside the space. De-energize the system using standard lockout/blockout procedures when opening a cover to a pressurized system. Square covers (i.e., grates) can fall into the opening. Take precautions to avoid dropping them by using a lifting device or by using another worker, especially if the cover is heavy, large, or awkward to handle alone.	Refer to Chapter 4.2 and 4.3 for additional information.* See 29 CFR 1910.22 and 29 CFR 1910.23 for regulatory details.
Respiratory hazards	Confined space entry; working with chemicals; biosolids dust, welding fumes, dry polymers, and other particulates; and emergency response to chemical releases.	Toxic and/or oxygen-deficient atmospheric conditions, acute or chronic respiratory hazards from inhalation of toxic chemical vapors or offgases, respiratory irritation from dry particulates, and high concentration of toxic gases when responding to a chemical release.	Test the air before entering any confined space. If the atmosphere is immediately dangerous to life or health (IDLH), a full-face pressure-demand SCBA with a minimum service life of 30 minutes or a combination full-face, pressure-demand, supplied-air respirator with an auxiliary self-contained air supply must be worn. When working with chemicals, review the label and/or SDS to determine if a respirator is required and/or recommended. Wear a particulate-type respirator if protection is needed when working with dusts or other dry, nonhazardous products. Wear an SCBA when responding to emergency chemical releases, especially if the concentration is unknown or if the chemical is offgassing.	Refer to Chapter 10 for additional information.* See 29 CFR 1910.134 and CFR 1910.146 for regulatory details.
Sampling and working over/near water	Sampling from basins, channels, and other treatment processes; sampling from streams, canals, and other moving bodies of water; sampling from lakes, ponds, lagoons, and other still waters; sampling from remote effluent structures; and sampling from watercraft.	Drowning exposures from falling into basins; standing in moving waters, or when sampling from watercraft; soft tissue injuries when carrying heavy sampling equipment; rashes and allergic reactions to poisonous vegetation, snakes, bees, and spiders when sampling from highly vegetated locations; and slip/trip hazards when carrying sampling equipment over rough terrain or when traversing up and down steeply sloped earthen or concrete embankments.	Do not climb on, climb over, or lean over guardrails or other protective barriers. If it is necessary to extend beyond a protective fall protection barrier, use personal fall arrest or positioning equipment. Wear a U.S. Coast Guard-approved personal flotation device (PFD) when working on or in water, especially if working over deep water, near/in fast-moving water, or if you cannot swim. Wear a U.S. Coast Guard-approved PFD if there is a chance you may be "pulled in" while performing grab sampling. Employees who will use watercraft to collect samples must be trained on boating safety and have a valid U.S. Coast Guard license for the class of watercraft and type of water being navigated. Use good lifting techniques when lifting automatic samplers. If the sampler is full or too heavy, use a jib crane, mechanical hoist, or have another person help lift it. Practice good landscaping around remote sampling sites that are subject to overgrowth of vegetation to discourage poisonous plants and animals. Use a sturdy walking stick or have a secure anchor point when traversing up and down steep embankments.	Refer to Sections 2.2 through 2.4 for additional information.* See 20 CFR 1926.106 for regulatory details.
Slip/trip hazards	Wet surfaces; chemical spills; poor housekeeping; loose grates, covers, uneven surfaces; traversing down ramps; stairwells; and sloped surfaces.	Contusions, broken bones, and miscellaneous injuries.	Immediately clean up spills to floors or walkways or use a cone or other warning device to alert people to slippery conditions. Clean up all chemical spills immediately. Secure all grates or covers or replace them if not flush with the pavement or walkway to prevent trip hazards. Wear sturdy work boots with good tread. Use handrails or other holds when traversing down slopes or stairways.	Refer to Sections 2.2 through 2.5 for additional information.*
Traffic hazards	Setting up equipment, entering confined spaces, and cleaning or inspecting sewers or other activities performed on public roads.	Fatalities and serious injuries from vehicular incidents.	Set up work zones according to U.S. Department of Transportation's *Manual on Uniform Traffic Control Devices (MUTCD) for Streets and Highways*.	See CFR 23, Part 655, Subpart F for regulatory details.

* For Chapter and Section references see: *Safety, Health, and Security in Wastewater Systems*.

Water Environment Federation
the water quality people®
www.wef.org

can be present in large quantities in the collection system. Bacterial infections, therefore, will result from their proliferation in an aqueous environment. Table 1.2 provides a summary of the various diseases associated with wastewater-contaminated environments. Many of these bacteria are typically transmitted via the fecal-oral transmission route (i.e., through oral ingestion of contaminated food and water or through hand-to-mouth contact).

Because of their daily exposure to wastewater-contaminated environments, wastewater personnel have a higher incidence of potential exposure to pathogens than the general public. For most workers, however, the risk of developing a disease is relatively low. However, antimicrobial resistance is an emerging concern. The practice of good personal hygiene and personal protective equipment (PPE) are critical because infections, including those resistant to antibiotics, may occur without symptoms. Asymptomatic workers can then spread the microbes to their family and members of the community without ever experiencing an illness (latent infection).

The most common bacterial pathogens found in wastewater are *Salmonella* and *Shigella*. Other bacterial pathogens include *Vibrio, Clostridium, Yersinia, Campylobacter, and Leptospira. Escherichia coli (E. coli)*, are generally considered non-pathogenic, but disease-causing strains do exist (e.g., Shiga toxin-producing *E. coli*, Enterotoxigenic *E. coli*, Enteropathogenic *E. coli*). The most common bacterial pathogens found in wastewater are listed in Table 1.2.

1.1.1 Salmonella

Salmonella is a significant cause of food poisoning from improperly prepared products. *Salmonella* can cause infections of the stomach and intestinal tract (acute gastroenteritis), typhoid fever, and paratyphoid fever. *Salmonella* infection results from oral ingestion, although large numbers of these microorganisms are required to cause illness. *Salmonella* are routinely isolated from raw and partially treated wastewater, compost operations, sludge handling facilities, and associated landfills.

1.1.2 Shigella

Shigella (predominately *Shigella sonnei* and *Shigella flexneri*) causes about 450 000 cases of diarrhea annually in the U.S. (CDC, 2019). *Shigella* is typically transmitted through ingestion (the fecal-oral transmission route). However, *Shigella* survives for only a short time in the collection system,

TABLE 1.2   Bacterial pathogens found in wastewater.

Microorganism	Disease
E. coli[a]	Gastroenteritis, Hemolytic uremic syndrome
Salmonella	Salmonellosis (gastroenteritis), Typhoid fever
Shigella	Shigellosis (gastroenteritis), Bacillary dysentery
Vibrio	Cholera
Clostridium	Tetanus, Gas gangrene, Gastroenteritis
Yersinia	Acute gastroenteritis
Campylobacter	Acute bacterial enteritis
Leptospira	Weil's disease (Leptospirosis)
Legionella	Legionnaires' disease

[a]Many *E. coli* strains are non-pathogenic. Pathogenic *E. coli* include Shiga toxin producing *E. coli*, Enterotoxigenic *E. coli* (ETEC), Enteropathogenic *E. coli* (EPEC).

and generally represents a greater potential hazard for collection system workers than treatment facility operators.

1.1.3 Vibrio

Cholera is caused by *Vibrio cholerae*, which produce a toxin that results in vomiting, diarrhea, and loss of body fluids. Cholera can be spread by the ingestion of fecally contaminated water and is typically present in many developing countries and communities with inadequate sanitation practices. Control of this disease is achieved through proper measures such as water disinfection and wastewater treatment.

1.1.4 Clostridium

Clostridium is a spore-forming bacterium. *Clostridium perfringens* can be found in soil, food, sewage, and the gastrointestinal (GI) tract of both diseased and non-diseased humans and animals. Frequently associated with foodborne illness, it can cause significant systemic and enteric diseases, including gas gangrene, gastroenteritis, and enterocolitis in both animals and humans (Kiu & Hall, 2018). *Clostridium tetani* causes tetanus, which may result from a localized infection of a deep or puncture wound. Symptoms of infection include contraction of the muscles controlling the jaw, body muscle spasms, and paralysis of the throat muscle, which can lead to death from respiratory failure. Infection may occur whenever a deep wound is contaminated with wastewater-contaminated material. The general public, including wastewater system personnel, should make sure that tetanus vaccines are taken every 10 years after initial doses and after wounds, unless it has been fewer than 5 years since the last dose. A booster tetanus toxoid given at the time of injury will also provide immunity to the disease.

1.1.5 Yersinia

Yersinia entercolitica is an enteric pathogen that causes acute gastroenteritis. The most common symptoms are fever and diarrhea, with moderate dehydration. The fecal-oral route is the most common mode of transmission.

1.1.6 Campylobacter

Campylobacter fetus and *C. jejuni* cause acute bacterial enteritis (Tortora et al., 1982). These organisms are transmitted by the fecal-oral route, and most outbreaks of enteritis caused by *Campylobacter* have been associated with poultry, raw milk, and untreated water (CDC, 2020b).

1.1.7 Leptospira

Leptospira bacteria, spread through the urine, are responsible for leptospirosis, or Weil's disease, which infects the liver, kidneys, and central nervous system. This disease was known as "the illness of the wastewater worker" in England. The Centers for Disease Control and Prevention currently lists *Leptospira* as an occupational hazard for sewer workers (CDC, 2020c). Infection typically occurs by way of contact with mucous membranes or skin abrasions (CDC, 2020c).

1.1.8 Legionella

Legionella causes Legionnaires' disease, a severe lung infection caused by breathing in small droplets of water that contain *Legionella* bacteria. Persons aged ≥50 years, current or former smokers, and those with chronic diseases or a weakened immune system are at higher risk for Legionnaires' disease (Soda et al., 2017). Wastewater treatment plants may play a role in local and community cases and outbreaks of Legionnaires' disease. Specifically, aerobic biological systems provide an optimum environment for the growth of *Legionella* due to high organic nitrogen and oxygen concentrations, ideal temperatures, and the presence of protozoa (Caicedo et al., 2019).

1.2   Viruses

A virus is any of a group of ultramicroscopic agents that reproduce only in living cells. This characteristic of viruses is important because viruses cannot reproduce without a host cell and, therefore, will not reproduce in wastewater. A primary source of viruses that are infectious to humans is human waste discharged to the sewer.

Over 100 different types of viruses are found in human waste. Human viruses commonly found in wastewater are listed in Table 1.3 (WEF, 2001). The majority of these viruses multiply in the living cells of the intestinal tract and end up in human feces. Because millions of viruses can be produced by an infected cell, they are found in large numbers in wastewater. Characteristics of various wastewater viruses are in Table 1.4 (WEF, 2001).

While there are many types of viruses present in wastewater, the general cate- gory that has received the most study is the enteric, or intestinal, virus. This group includes varieties that are responsible for diseases such as: infectious hepatitis, meningitis, poliomyelitis, respiratory diseases, gastroenteritis, and the common cold. Additionally, enteric viruses of concern (Table 1.3) are non-enveloped viruses, consisting of a protein capsid and nucleic acid (RNA or DNA), and are more persistent in wastewater and resistant to some disinfectants (such as chloramines), as compared to enveloped viruses. Many of these enteric viruses are also infectious at very low doses. Thus,

TABLE 1.3   Human viruses generally found in wastewater.

Virus group	Disease
Norovirus	Acute gastroenteritis
Rotavirus	Acute gastroenteritis
Adenovirus	Acute respiratory disease, conjunctivitis, pharyngoconjunctival fever
Coxsackie A	Upper respiratory tract infection
Coxsackie B	Upper respiratory tract infection, myocarditis, aseptic meningitis, Bornholm's disease
Echovirus	Common cold, aseptic meningitis, conjunctivitis, gastroenteritis
Hepatitis A	Infectious hepatitis
Poliovirus	Poliomyelitis
Reovirus	Upper respiratory tract infection

TABLE 1.4   Characteristics of various viruses found in wastewater.

Virus group	Mode of transmission	Incubation period
Adenovirus	Ingestion or inhalation	5–7 days
Echovirus	Inhalation	1–2 days
Hepatitis A	Ingestion	15–40 days
Poliovirus	Ingestion	5–20 days
Norovirus	Ingestion	1–2 days
Rotavirus	Ingestion	
Coxsackie A	Ingestion or inhalation	
Coxsackie B	Ingestion or inhalation	

wastewater personnel have a higher potential incidence of exposure to these viruses because of their daily contact with wastewater. On the other hand, enveloped viruses (e.g., coronavirus, influenza, and HIV), which consist of a lipid membrane, protein capsid, and nucleic acid, are more sensitive to environmental stresses and wastewater treatment processes. Once their lipid envelope is disrupted, the virus is no longer infectious or able to replicate. Thus, enveloped viruses are generally less of a concern to people in contact with wastewater.

1.2.1   Hepatitis A Virus

The main waterborne disease resulting from viral infection is hepatitis A. The hepatitis A virus is the causative agent of infectious hepatitis, a systemic disease primarily involving the liver. The virus is commonly associated with fecal-oral transmission through wastewater contamination and contaminated food. An infected person generally exhibits flu-like symptoms, cramps, vomiting, high fever, and jaundice.

1.2.2   Norovirus

Norovirus is another common type of virus associated with inadequately treated wastewater, which causes acute gastrointestinal disease consisting of vomiting, diarrhea, low-grade fever, and body aches. Symptoms generally last for a short period of time, typically 24 to 48 hours. During this time, the virus can be passed through the stool and has the potential to affect other members of the family if appropriate hygiene is not practiced. Norovirus occurs in high densities in raw wastewater (Eftim et al., 2017), and outbreaks of the illness have been associated with wastewater disposal, municipal water supplies, and recreational water contact.

1.2.3 Adenovirus

Adenoviruses have been associated with respiratory tract infections, conjunctivitis (eye infection), and acute viral gastroenteritis. The virus is commonly isolated from wastewater and sludges at high densities.

1.2.4 Rotavirus

Rotaviruses are a common cause of acute viral gastroenteritis. Raw wastewater and some chlorinated wastewater effluents from activated sludge facilities treating domestic wastes have been shown to discharge high densities of these viruses.

1.2.5 Coxsackieviruses A and B

Coxsackievirus A causes aseptic meningitis and conjunctivitis and is one of the causes of the common cold. Coxsackievirus B causes several types of diseases, including heart disease. The primary modes of transmission for coxsackieviruses are through inhalation or ingestion of contaminated materials.

1.2.6 Poliovirus

The poliovirus is associated with poliomyelitis, which affects the central nervous system. The primary mode of transmission is ingestion of fecally contaminated water containing the virus. Poliovirus vaccines have reduced the incidence of poliomyelitis and have contributed to the decline in reported cases of the disease. Outbreaks typically occur only in segments of the population lacking proper immunization.

1.2.7 Human Immunodeficiency Virus (HIV)

Acquired immune deficiency syndrome (AIDS) is caused by the human immunodeficiency virus (HIV) that attacks the body's immune system, leaving the body susceptible to numerous diseases. HIV is an enveloped virus that cannot survive for long periods of time outside of the human body. The Centers for Disease Control and Prevention (CDC) has stated that there is no evidence that HIV is spread in wastewater or its aerosols. Studies have shown that HIV survives poorly in wastewater with a 1-3 log reduction in 24–28 hours (Casson et al., 1992). There have been no known cases of wastewater workers or plumbers who have contracted AIDS where the mode of transmission was judged to be from occupational exposure.

1.2.8 Severe Acute Respiratory Syndrome Coronavirus 2 (SARS-CoV-2)

Severe acute respiratory syndrome coronavirus 2 (SARS-CoV-2) is the virus that causes the coronavirus disease (COVID-19). People with COVID-19 experience a wide range of symptoms—from mild to severe. Common symptoms include, but are not limited to, dry cough, fever, shortness of breath, chills, muscle pain, sore throat, and a loss of smell or taste. Some patients have also experienced gastrointestinal symptoms, such as diarrhea and vomiting (CDC, 2020a).

The main routes of transmission are respiratory droplets and direct contact, with no reports of fecal-oral transmission to date. Additionally, SARS-CoV-2 is an enveloped virus and less stable in the environment compared to many non-enveloped enteric viruses found in untreated wastewater. Standard treatment and disinfectant processes at water resource recovery facilities (WRRFs) are expected to be effective at inactivating SARS-CoV-2. While genetic material of the virus has been detected in untreated wastewater (Ahmed et al., 2020), evidence is not definitive that the virus is infectious in stool or sewage. One study has, however, indicated SARS-CoV-2 may be infectious in urine (Sun et al., 2020).

Overall, the risk of transmission of SARS-CoV-2 from feces of an infected person or via sewage systems appears to be low. Best practices for protecting the health of workers exposed to wastewater should be followed, such as: engineering and administrative controls; safe work practices; and personal protective equipment normally required for work tasks when handling untreated wastewater. No additional COVID-19–specific protections are recommended for employees involved in wastewater management operations (Environmental Protection Agency [U.S. EPA], 2020a).

1.3 Parasites

A parasite lives on or in another organism of a different species, from which it derives its nourishment. The organism is called the parasite's *host*. Parasites typically do not kill their hosts because the life of the parasite would also be terminated.

In many instances, however, parasites will weaken the host or cause symptoms similar to disease caused by bacteria or viruses. Waterborne parasites found in wastewater consist of various types of protozoa and helminths (worms). Many of the life-stages of these organisms often do not survive the journey through the wastewater collection system and treatment facilities. However, the oocysts, cysts, and eggs, in which the protozoa and worms reproduce, are often resistant to adverse conditions, and therefore, may be present in wastewater or sludge samples.

The number and variety of parasitic forms present in wastewater or sludges depend heavily on the origin of wastes entering the treatment plant. The most commonly studied protozoa are *Entamoeba histolytica, Cryptosporidium parvum,* and *Giardia lamblia. E. histolytica* is the agent that causes amoebic dysentery, a disease with symptoms that include varying degrees of abdominal cramps and diarrhea, alternated with constipation. *G. lamblia* is also contracted orally and can lead to a variety of intestinal symptoms. *Giardia* is a hardy protozoon that exists in a cyst stage and can be resistant to chlorination. *Cryptosporidium parvum* can be found in human and animal feces and the oocysts are extremely resistant to chlorine disinfection. The most common parasites found in wastewater are listed in Table 1.5.

Additionally, the eggs of many varieties of helminths (e.g., roundworms, hookworms, and tapeworms) have also been found in wastewater. Infestation of roundworms and tapeworms is typically transmitted orally and typically results in abdominal pain and weight loss. Hookworms are generally transmitted through cracks in bare skin (such as between the toes), although oral infestation is also possible. Hookworms cause a general loss of energy and anemia.

Parasite survival rates are affected by the wastewater or sludge treatment processes to which they are subjected. In general, each process that exposes a parasite to a different or hostile environment may shorten its survival time. In instances of parasitic infestation, it is possible that the host's symptoms may be nonexistent. Because hand-to-mouth contact is the principal cause of infection, it is important to use appropriate PPE and to wash hands frequently.

TABLE 1.5   Parasites found in wastewater.

Organism	Disease
Protozoa	
Entamoeba histolytica	Amoebic dysentery
Giardia lamblia	Giardiasis
Cryptosporidium parvum	Cryptosporidiosis
Helminths	
Roundworms (nematodes)	Abdominal pain and weight loss
Hookworms (ancylostomatodes)	Anemia
Tapeworms (cestodes)	Abdominal pain and weight loss

1.4   Fungi

Fungi are diverse, ubiquitous in the environment, and some species may cause allergic reactions or affect those with compromised immune systems. Fungal spores, such as *Aspergillus* spp., have been found in bioaerosols associated with WRRFs and biosolids (Niazi et al., 2015). People with weakened immune systems or lung diseases are at a higher risk of developing health problems due to *Aspergillus*. Within WRRFs, the highest emission of bioaerosols occurs in pretreatment and primary clarifiers units and those containing moving mechanical equipment for wastewater aeration (Pascual et al., 2003). Given the size of fungal spores (1–30 µm), respiratory protection should be used to protect workers from fungal spore inhalation.

1.5   Macroorganisms

There are numerous places where rodents can be attracted to wastewater facilities. The most common areas where this occurs are screening, grit collection, and sludge treatment and disposal areas. Screenings and grit should be collected and stored in containers that minimize rodents from entering and congregating. General good housekeeping practices should also minimize rodents from becoming a problem.

Nematodes and bristle worms are the most common worms found in wastewater systems. Bristle worms can sometimes indicate high nitrates and turn the entire system pink (Environmental Leverage Inc., 2003).

Insects can also become a problem at WRRFs. Any areas of standing water or ponding can become breeding grounds for insects such as mosquitoes. Standing water, therefore, should be eliminated. If possible, tanks that contain water or wastewater and are not being used should be drained. Good housekeeping practices should be followed to prevent attraction of insects.

2.0   HOW INFECTIONS CAN SPREAD

In the wastewater setting, there are four basic routes that may lead to infection: 1) *ingestion* through wastewater splashes, contaminated food or beverages, or from pathogens on contaminated hands; 2) *inhalation* of infectious agents in aerosols or bioaerosols emitted during various wastewater processes or from close person-to-person contact; 3) *injection* through a cut or abrasion; and 4) and through skin contact. Wastewater workers often come in physical contact with raw wastewater and sludge during their daily activities. Even when direct physical wastewater contact is avoided, the worker may handle contaminated objects (fomites) and accidentally ingest pathogens via hand-to-mouth transmission. Cuts and abrasions, including those that are minor, should be cared for properly. Open wounds invite infection from many of the viruses and bacteria present in wastewater. Table 1.6 summarizes significant routes of infection.

Ingestion is generally the primary route of wastewater worker infection. The common practice of touching the face or mouth with the hand will contribute to the possibility of infection. Workers who eat or smoke without washing their hands have a much higher risk of infection. Most surfaces near wastewater equipment are likely to be covered with bacteria or viruses. These potentially infectious agents may be deposited on surfaces in the form of an aerosol or may come from direct contact with the wastewater or sludge. A good rule of thumb for a worker to follow is to never touch oneself above the neck whenever there is contact with wastewater. Table 1.7 lists methods to prevent ingestion of pathogenic organisms.

At locations where wastewater or sludge is sprayed, the possibility of inhaling infectious agents will increase. Workers should use appropriate PPE to prevent inhalation in areas where contact with such aerosols are likely. In instances where exposure to aerosols is anticipated, use of N95 respirators or surgical masks and goggles may help to minimize contact.

3.0   HOW TO PREVENT INFECTIONS

3.1   Work Procedures

The General Duty Clause of the Occupational Safety and Health Act (OSHA) states: 29 U.S.C. § 654, 51: "Each employer shall furnish to each of their employees employment and a place of employment which are free from recognized hazards that are causing or are likely to cause death or serious physical harm to their employees." Based on the prior discussion in section 1.0, it is clear that untreated and partially treated wastewater and biosolids

TABLE 1.6   Routes of infection.

Ingestion	Eating, drinking, or accidentally swallowing a pathogenic organism (e.g., hepatitis A).
Inhalation	Breathing aerosols or dust containing pathogenic organisms (e.g., common cold, *Legionella*, Coronaviruses).
Injection	Entry into the body via a cut, abrasion, or intravenous delivery.
Skin contact	Entry of pathogenic organism to body via skin contact.

TABLE 1.7   Methods to prevent ingestion of pathogenic organisms.

Proper Hygiene

    Never eat, drink, or use tobacco products before washing hands;

    Avoid touching face, mouth, eyes, or nose before washing hands; and

    Wash hands immediately after any contact with wastewater or sludge.

    Wash hands for at least 20 seconds using an antibacterial soap.

    If handwashing is not immediately available, use a hand sanitizer that has at least 60% alcohol. Rub hands together to cover all surfaces of the skin.

Control activities

    Eat only in designated areas of the plant and away from treatment facilities, and

    Do not smoke or use chewing tobacco while working in direct contact with wastewater or sludge.

present biological hazards to wastewater workers. It is therefore important that management 1) recognizes their responsibility under the General Duty Clause, 2) identifies workplace hazards, and 3) implements job-specific PPE and training on its proper use.

Because the specific biological hazard can vary by specific job activity and from facility to facility, it is necessary for facility managers and staff to conduct a Job Safety (or Hazard) Analysis (OSHA, 2002). A job safety analysis is a technique that focuses on job tasks as a way to identify hazards before they occur. The analysis identifies each task in a job, defines the potential hazards, and outlines critical safety practices. The analysis includes physical, chemical, biological, electrical, radiological, and gas/emissions. Each task-related hazard is ranked by probability, severity, and potential consequence. Once the hazards are known and prioritized, management and staff can then identify appropriate hazard control measures, which can include:

- Engineering controls
- Administrative controls
- Required PPE
- Required training
- Required permits
- Other information (e.g., vaccines, physical requirements, etc.)

3.2   Personal Protective Equipment

Within wastewater systems, biological hazards can be encountered through contact transfer, splashes, whole body contact, abrasions or cuts, and aerosols and dust. Table 1.8 summarizes recommended PPE based on the various routes of exposure. Table 1.9 demonstrates how the recommended PPE could be applied to various job activities.

Workers should remove PPE at the job site after the task has been completed. Workers should never wear PPE inside office areas, lunchrooms, or work vehicles as they can contaminate these areas. Training is critical not only how to put on and fit the appropriate PPE, but also on how to remove the PPE without contaminating the worker. When N95 respirators are used, training is necessary to ensure the proper fit and application (WHO, 2020). In some instances, surgical masks or dust masks may be sufficiently protective.

In addition to appropriate PPE, it is also important to practice proper personal hygiene to reduce the possibility of biological contamination, as outlined in Table 1.10. Workers should thoroughly clean up at the end of the workday before going home. All tools contaminated with wastewater should be cleaned with a common cleaner or a mild solution of sodium hypochlorite. First aid kits should be readily available at the jobsite to allow for the immediate treatment of minor cuts.

Soiled clothing should be laundered before it is taken home, or it should be handled separately from domestic laundry and washed with hot water and disinfected. Dual locker systems are desirable for all wastewater workers, allowing one locker for work clothes and one for street clothes.

TABLE 1.8   Recommended PPE based on the route of exposure (LeChevallier et al., 2019).

Route of exposure	Recommended PPE
Contact transfer	Gloves, boots, and uniform/coverall
Splash	Protection of the eyes/face through safety glasses, face shield, or goggles
Whole-body contact	Tyvek suits or coveralls
Abrasion, cut, or puncture	Durable gloves designed for protection from cuts or punctures
Respiratory from sprays, mists, or dust	N95 respirator, surgical or dust mask

TABLE 1.9 Example job activities and suggested personal protective equipment to protect wastewater workers (LeChevallier et al., 2019).

Area/Location	Activity	Contact Transfer	Splash (Eyes/Face)	Whole Body Contact	Abrasion, Cut, Puncture	Respiratory
Collection System	Lift Station Inspection	X				
	Vacuum/Jetter Truck Operation	X	X			C
	Netting Facility/Storm Drain Pretreatment O&M	X	X		X	
	CCTV or Line Cleaning	X	X		X	
	Field Wastewater Sampling	X				
	Sewer Entry (Live)	X	X	X	X	X
	Sewer Entry (By-pass)	X	X		X	
	Man-hole Maintenance	X			X	
	Sewer Pipe Repair Work (Live)	X	X	X	X	X
	Sewer Pipe Repair Work (By-pass)	X	X		X	
	Spill Response/SSO/CSO	X	X		X	
Routine Facility Operator Activities	Visual Process/Plant Inspections	X				
	Pushbutton Equipment Operation	X				
	Manual Valve Operation	X			X	
	WW Sample Collection (Auto)	X				
	WW Sample Collection (Grab)	X				
	Field Instrument Calibration (e.g., Dissolved Oxygen Meter)	X	X			
	Sludge Judge	X	X			
	General Housekeeping (Hose-down)	X	X			
	Dry Sweeping, High-Pressure Power Wash	X	X			C
	Lab Activities	X	X		X	
	Hand-Held Dissolved Oxygen Meter	X	X			
Maintenance	Facility Maintenance/Daily Rounds	X				
	Active Pump and Line Maintenance	X	X	X	X	C
	Process and Equipment Maintenance with Sewage Contact	X	X			
	Tank Entry (Empty Tank) Maintenance Activities	X	X	X		C
Preliminary Equipment	Cleaning Bar Screens	X	X		C	C
	Screenings Handling	X	C		C	
	Grit Handling	X	C			
UV Disinfection	Routine Inspection	X				
	Routine Maintenance	X				
	Bulb Replacement	X	X		X	
	Ballast Replacement	X	X			
Biosolids Handling Processes	Gravity Thickening Operation	X				
	Other Thickening (Dissolved-Air Flotation, Gravity Belt Thickener, Drum) Operations	X	X			
	Open Dewatering Equipment Operation	X	X			
	Enclosed Dewatering Equipment Operation	X				
	Liquid & Cake Sampling	X				
	Septage/Waste Receiving	X	X			C
	Compost Handling	X				C
	Dewatered Class B Biosolids Handling	X				C
	Dewatered Class A Biosolids Handling	X				
	Thermally Dried Biosolids/Ash Handling	X				

NOTES: X = recommended; C = conditional depending on specifics of the task. See Table 1.8 for description of recommended PPE based on route of exposure.

TABLE 1.10   Workplace precautions and personal hygiene guidelines.

- Remove PPE (gloves, coveralls, etc.) before entering office areas, lunchrooms, or work vehicles.
- Wash hands frequently with soap and water after contacting wastewater; visiting restrooms; before eating, drinking, or smoking; and at end of work shift.
- Promptly treat cuts and abrasions using appropriate first aid measures.
- Change soiled uniforms or protective clothing as soon as the job is completed.
- Shower before changing into clean work clothes and shoes.
- Launder work clothes at work not at home.
- Handle sharp items with extra care to prevent accidental injuries.
- Clean contaminated tools after use.
- Exercise extra caution whenever there is contact with contaminated water or sludge.
- Wherever possible, use dual lockers to separate work and street clothes.
- Promptly clean body parts that contact wastewater or sludges.

3.3   Immunizations

The CDC recommends that a vaccination program for workers exposed to wastewater or human waste be developed in consultation with local health authorities (CDC, 2020d). The CDC recommends that tetanus vaccinations should be up to date, with consideration also given to the need for polio, typhoid fever, hepatitis A, and hepatitis B vaccinations. The tetanus booster should be repeated if a wound or puncture becomes dirty and if boosters have not been given within 10 years. The CDC maintains a list of recommended vaccinations for children, adults, and health care workers (CDC, 2020e).

3.4   Personal Protection Measures

As discussed previously, when working at WRRFs, the most important practice is personal hygiene and hand washing. Workers should avoid direct contact with wastewater by wearing gloves, boots, water resistant coveralls, eye and face protection, and other job-specific PPE. In short, workers should assess the risk of exposure to wastewater and wear the proper PPE. When a cut or abrasion does occur, the worker should seek medical attention as soon as possible to clean and dress the affected area.

4.0 HOW TO TREAT INFECTIONS

First and foremost, if an employee experiences symptoms of illness, such as fever, coughing, shortness of breath, dizziness, nausea, etc., the employee should not enter the workplace but remain at home to avoid exposing others to potential infectious illness. If a worker is injured, all injuries should be reported and treated promptly to prevent infection or illness. Table 1.11 lists the minimum recommended contents of a standard first aid kit. Depending on the degree of medical training of the staff, additional items can be included (e.g., automated external defibrillator). Potential entry points will exist for microorganisms to cause infection if minor breaks in the skin or mucous membranes resulting from burns, rashes, cuts, and insect bites are left untreated. Soap and water are the best initial first aid measures that can be used for minor cuts. An antibiotic ointment or disinfectant should be applied to any wound after thorough washing. Adhesive bandages, tape, and sterile gauze should also be used to further protect the treated area and to keep the wound clean and dry. Generally, prompt medical attention is required if injuries result from contact with contaminated wastewater or sludge or if wounds or punctures do not respond to methods to control bleeding. The placement of first aid kits in a WRRF should be determined based on an area's acceptability for applying prompt medical attention. As a general rule, break rooms, maintenance shops, and operation centers are acceptable.

TABLE 1.11 Suggested contents of a first aid kit.

Anti-bacterial ointments and/or creams/disinfectants
Dressings
Waterless soap
Antiseptic wipes
Hand sanitizer
Latex gloves
Surgical mask
Sterile eyewash solution
Adhesive bandages
Tape
Scissors and tweezers
Sterile gauze
Splint
Allergy medications
Analgesic

Because wastewater workers frequently expose their hands to water and wastewater, occasional skin problems such as fungal infections, rashes, chapping, and cracking may occur. Protective hand creams or lotions can typically be used to minimize such problems. If these medications are ineffective or if contact dermatitis becomes a problem, an occupational physician should be consulted. If wastewater gets splashed into the eyes, ears, or nose of a worker, it should be immediately flushed with fresh potable water or a sterile solution from the first aid kit. For areas where splashing of wastewater can occur, such that water droplets are aerosolized, personnel should always wear the appropriate PPE to guard against potential exposure to biological or chemical hazards that may be present.

5.0   WORKERS WHO ARE AT RISK

Several studies have been conducted on the actual infection rate of wastewater workers (Carducci et al., 2000; Carducci et al., 2016; Lin and Marr, 2017; Thorn & Kerekes, 2001). During early years of employment, wastewater workers may be more prone to illness than more experienced workers. Newer employees may experience increased rates of gastrointestinal and upper respiratory illnesses, which are thought to be related to biological exposures (Thorn & Kerekes, 2001). Table 1.12 shows the results of various

TABLE 1.12   Summary of risks to wastewater and biosolids workers.

Type of hazard	Effects observed
Hepatitis A	Evidence of increased risk when working with raw wastewater and primary sludge (Thorn & Kerekes, 2001).
Other viral infections	May indicate infection in the most exposed workers. Other factors contributing to infection should not be overlooked (Carducci et al., 2018).
Leptospirosis	Infections were significantly higher among wastewater treatment workers than in a comparison group (Al-Batanony & El-Shafie, 2011).
Gastrointestinal illness	Increased rates, especially among new workers (Thorn & Kerekes, 2001).
Biosolid dust	Excess nasal, ear, and skin abnormalities and eye irritation (Oppliger et al., 2005).

studies of health effects of biological hazards to wastewater workers. Simple procedures involving training, appropriate PPE, personal hygiene, and work methods, however, can reduce these risks to below other common occupational hazards.

Most studies have indicated that areas with the greatest risk for infection involve routine and direct contact with untreated wastewater or sludge (Al-Batanony & El-Shafie, 2011; Oppliger et al., 2005). Included in this category are workers involved in sewer maintenance and raw sludge handling. Various treatment processes designed for solids or biochemical oxygen demand removal will provide varying degrees of pathogen removal, as shown in Table 1.13. Risk of exposure should decline as wastewater undergoes various treatment steps (Metcalf and Eddy, 2013).

5.1   Collection System Personnel

Because of their direct high exposure to raw wastewater, collection system workers have greater risks of infection, as compared to treatment facility personnel. Although various studies have indicated evidence of increased risk of viral infections (including hepatitis A), parasite infection, and gastrointestinal illness in these workers (Thorn & Kerekes, 2001), application of appropriate PPE and hygiene practices can reduce these risks. Collection system personnel must be aware of not only contact risk with raw wastewater, but also consider splash, whole body exposure, and the possibility of aerosol inhalation. Activities like cleaning a clogged sewer main can create aerosols that pose risks not only to the worker but also potentially to the public; considerations should be made to apply appropriate setbacks on the work zone to minimize any exposure to the public.

5.2   Treatment Facility and Laboratory Personnel

Studies have shown increased risk for operators involved with raw sludge handling or in enclosed areas where wastes are aerated or agitated (Oppliger et al., 2005). Laboratory personnel are required to perform analyses on a variety of wastewater and sludge samples. Although the risk of infection from wastewater samples is not as high in the laboratory environment as in collection systems or outside facilities, infectious agents that are commonly found in such samples are nevertheless a biological hazard. The risk of laboratory-acquired infection results from any procedure that releases infective organisms to the environment or affords access for such organisms to the human body. The laboratory environment, therefore, should be properly ventilated by a plenum and exhaust air system to minimize exposure to chemical or biological risks. Exhaust air should be routed from the laboratory to the outside of the building and discharged to the atmosphere.

TABLE 1.13 Pathogen $\log_{10}$ reduction ranges across unit treatment processes (adapted from Soller et al., 2018).[a]

	Adenovirus	Campylobacter	Cryptosporidium	Giardia	Norovirus	Salmonella
CSWT[b]	0.9–3.2	0.6–2.0	0.7–1.5	0.5–3.3	0.8–3.7	1.3–1.7
Ozonation	4.0	4.0	1.0	3.0	5.4	4.0
BAF	0–0.6	0.5–2.0	0.0–0.9	0.0–3.9	0.0–1.0	0.5–2.0
MF	2.4–4.9	3.0–9.0	4.0–7.0	4.0–7.0	1.5–3.3	3.0–9.0
RO	2.7–6.5	4.0	2.7–6.5	2.7–6.5	2.7–6.5	4.0
UF	4.9	5.6–9.0	4.4–6.0	4.7–7.4	4.5	5.6–9.0
UV Dose						
800 mJ/cm^2	6.0	6.0	6.0	6.0	6.0	6.0
12 mJ/cm^2	0.0–0.5	4.0	2.0–3.5	2.0–3.5	0.5–1.5	4.0
Chloramines[c,d]	1.0–4.0	1.0–4.0	1.0–3.0	1.0–3.0	1.0–4.0	1.0–4.0

[a]log10 units; Adenovirus IU/L, *Campylobacter* MPN/L, *Cryptosporidium* oocysts/L, *Giardia* cysts/L, Norovirus copies/L, *Salmonella* PFU/L.

[b]CSWT = conventional secondary wastewater treatment; BAF = biologically active filtration; RO = reverse osmosis; UF = ultrafiltration; UV = ultraviolet radiation.

[c]Chloramines (or combined chlorine) are more typical of wastewater treatment, given the presence of ammonia in wastewater. Reductions vary with applied CT. The CT value is the product of the disinfectant concentration (C) and the contact time (T) with the water being disinfected. The CT value for a specific disinfectant varies based on specific temperatures and pH (U.S. EPA, 1991).

[d]Ballester and Malley (2004); Metcalf and Eddy (2007); Cromeans et al. (2010).

Laboratory workers must be provided with adequate training in proper microbiological techniques and safety. Workers should be familiar with aseptic handling techniques and the biology of the organisms under evaluation to fully appreciate the potential hazard. An emergency procedure should also be developed to deal with accidental contamination of personnel and work areas and appropriate vaccinations should be administered if known pathogens are being evaluated. All laboratory apparatuses and waste should also be decontaminated by disinfection or sterilization to keep the environment free from microorganisms. Table 1.14 lists recommended safe laboratory practices.

5.3 Biosolids Personnel

Wastewater solids can comprise a range of materials, from raw sludges, to treated biosolids, which, depending on the level of pathogen reduction can be categorized as either Class A or Class B biosolids (U.S. EPA, 2020b). Specific studies have been conducted on workers who deal with wastewater sludge composting (CDC, 2002). The heat generated in a properly managed composting operation is sufficient to significantly reduce levels of pathogens of concern in the wastewater industry. The conditions created in composting, however, allow for the proliferation of many thermophilic microorganisms such as *Aspergillus fumagatus*. *A. fumagatus* grows well at 45 °C (113 °F) and higher, which makes it prevalent at composting sites. The mode of infection is by way of inhalation of *A. fumagatus* spores in the dust at the site. Symptoms that have been reported by workers include abnormal skin, ear, and nose infections. Higher rates of eye and skin irritations have also been noted (Clark et al., 1984; Clark, 1987; Oppliger et al., 2005; Thorn & Kerekes, 2001). It is important, therefore, that appropriate eye and respiratory protective measures should be used. Use of dust masks may be

---

TABLE 1.14 Safe laboratory practices.

- Do not eat, drink, or smoke in the laboratory or while handling wastewater or sludge samples.
- Wash hands often while working in the laboratory.
- Wear protective clothing, laboratory coats, eye protection, and latex gloves, as required.
- Do not place hands on face, eyes, nose, or mouth while working in the laboratory; always keep your hands below the collar.
- Use bulb/pipette aid to pipette samples; do not pipette by mouth.
- Wipe up and disinfect spills immediately.
- Disinfect and discard all unused samples immediately.
- Take extra precautions when handling glassware to prevent breakage injury.

appropriate when cleaning biosolids handling areas and locations within the treatment facility where raw or partially treated wastewater has dried.

Compost handling is different from dewatered Class B biosolids handling, dewatered Class A biosolids handling, and thermally dried biosolids/ash handling, as outlined in 40 C.F.R. part 503. Class B biosolids are treated to achieve significant (i.e., 99%) pathogen reduction and are subject to site use and access restrictions. Class A biosolids (including thermally dried biosolids) are disinfected to a level that inactivates pathogens and are subject to fewer site-specific controls. Class A biosolids that also have sufficiently low heavy metal concentrations are referred to as Class A, EQ (exceptional quality) biosolids and can be bagged and distributed for home use.

Existing requirements and guidance help ensure that biosolids are processed, handled, and land applied in a manner that minimizes the risk of exposure to pathogens, including viruses. Generally, pathogens may exist when requirements are met under 40 C.F.R. part 503 for Class B biosolids, which is why U.S. EPA's site restrictions that allow time for pathogen degradation should be followed for harvesting crops and turf, for grazing of animals, and public contact. Additionally, per CDC's *Guidance for Controlling Potential Risks to Workers Exposed to Class B Biosolids* (CDC, 2002), employers should prevent work-related illness by providing proper PPE and supporting other health and safety practices for persons hauling and land applying biosolids. In addition to Class B safety precautions, 40 C.F.R. part 503 lists the stricter Class A biosolids pathogen requirements. To ensure employee safety with working with Class A biosolids, follow the safe laboratory practices outlined in the Clean Water Act 40 C.F.R. part 503 guidelines.

5.4   Other Personnel Who May Be Exposed to Fecal Contamination

5.4.1   Overflows and Stormwater

Combined sewers are designed to collect both sanitary sewage and stormwater in a single-pipe system. Overflows known as combined sewer overflows (CSOs) can occur during heavy rainfall or snowmelt and untreated stormwater and wastewater is discharged to a local waterbody. Separate sanitary sewers systems carry only wastewater but can also overflow during heavy rainfall or snowmelt due to infiltration and inflow. Municipal separate storm sewer systems carry only stormwater. Stormwater systems workers who maintain these systems can be exposed biological risks similar to collection systems workers, although generally under spill response conditions. Many times, these workers also conduct collection systems maintenance. Their job tasks (and risks) mirror those of collections systems workers.

5.4.2 Septic Systems Hauling and Maintenance

Workers in the septic system hauling and maintenance areas may have a high level of exposure through direct contact with untreated wastewater when performing tasks such as pumping, jetting, cleaning, transport, and repairs. Equipment handling and disinfection and other best practices have been identified (Washington On-site Sewage Association, 2020).

5.4.3 Landfill Leachate

Landfill leachate handling can be another form of exposure for wastewater workers, with splashing being the most likely route of exposure. A subcategory of landfill leachate is hazardous and medical waste exposure. Studies have shown the presence of bacterial, protozoan, and viral pathogens in municipal landfills (Gerba et al., 2011).

5.4.4 Industrial Pretreatment

The Clean Water Act Industrial Pretreatment Program requires continuous monitoring of industrial operations. Typically, wastewater from industrial operations does not contain pathogens; however, because hazards in industrial operations can vary, and pretreatment workers can handle sampling equipment and samples and can be in contact with wastewater through manholes, it is important to conduct a job safety analysis for each facility and take similar precautions to collection system workers.

5.4.5 Ambient Water Sampling

Human fecal contamination can enter ambient surface waters from combined or sanitary sewer overflows or spills. Therefore, appropriate precautions should be taken when sampling fecally contaminated waters. Some wastewater facilities have sampling programs beyond the boundaries of their property or outside the Industrial Pretreatment Program. These workers can handle samples from rivers, lakes, and coastal waters and should be aware of the potential for biological hazards and use the appropriate PPE during sample collection and analysis.

6.0 SUMMARY

A number of occupational hazards confront treatment plant and wastewater collection system workers. The danger of infection to these workers through contact with wastewater is real if proper safety precautions are not observed. Although the possibility of infection is greatest for workers in high-exposure areas, such as collection systems and raw sludge processing, all workers who handle or come in contact with wastewater are susceptible to infection.

The incidence of occupational illness or disease among experienced wastewater workers can be comparable to other non-wastewater-related professions provided proper training and PPE are utilized. Wastewater workers, however, must be alert to the potential for illness and should use common sense and follow safe work procedures. The implementation of strong safety programs, good personal hygiene practices, and PPE and clothing will minimize the risk of exposure to infectious agents commonly found in wastewater and sludge.

7.0   REFERENCES

Ahmed, W., Angel, N., Edson, J., Bibby, K., Bivins, A., O'Brien, J. W., Choi, P. M., Kitajima, M., Simpson, S. L., Li, J., Tscharke, B., Verhagen, R., Smith, W. J. M., Zaugg, J., Dierens, L., Hugenholtz, P., Thomas, K. V., & Mueller, J. F. (2020). First confirmed detection of SARS-CoV-2 in untreated wastewater in Australia: A proof of concept for the wastewater surveillance of COVID-19 in the community. *Science of the Total Environment*, 728. 10.1016/j.scitotenv.2020.138764

Al-Batanony, M. A., & El-Shafie, M. K. (2011). Work-related health effects among wastewater treatment plants workers. *International Journal of Occupational and Environmental Medicine, 2*(4), 237–244.

Ballester, N. A., & Malley, J. P. (2004). Sequential disinfection of adenovirus type 2 with UV-Chlorine-Chloramine. *Journal AWWA, 96*(10), 97–103. https://doi.org/10.1002/j.1551-8833.2004.tb10726.x

Caicedo, C., Rosenwinkel, K.-H., Exner, M., Verstraete, W., Suchenwirth, R., Hartemann, P., & Nogueira. R. (2019). *Legionella* occurrence in municipal and industrial wastewater treatment plants and risks of reclaimed wastewater reuse: Review. *Water Research, 149*, 21–34. https://doi.org/10.1016/j.watres.2018.10.080

Carducci, A., Tozzi, E., Rubulotta, E., Casini, B., Cantiani, L., Rovini, E., Muscillo, M., & Pacini, R. (2000). Assessing airborne biological hazard from urban wastewater treatment. *Water Research, 34*(4), 1173–1178. DOI: 10.1016/S0043-1354(99)00264-X

Carducci, A., Donzelli, G., Cioni, L., & Verani, M. (2016). Quantitative microbial risk assessment in occupational settings applied to the airborne human adenovirus infection. *International Journal of Environmental Research and Public Health, 13*(7), 733. https://doi.org/10.3390/ijerph13070733

Carducci, A., Donzelli, G., Cioni, L., Federigi, I., Lombardi, R., & Verani, M. (2018). Quantitative microbial risk assessment for workers exposed to bioaerosol in wastewater treatment plants aimed at the choice and setup of safety measures. *International Journal of Environmental Research and Public Health, 15*, 1490. doi:10.3390/ijerph15071490

Casson, L. W., Sorber, C. A., Palmer, R. H., Enrico, A., & Gupta, P. (1992). HIV survivability in wastewater. *Water Environment Research, 64*, 213–215. doi:10.2175/WER.64.3.4

Centers for Disease Control and Prevention. (2002). Guidance for Controlling Potential Risks to Workers Exposed to Class B Biosolids. National Institutes for Occupational Safety and Health: 2002-149. https://www.cdc.gov/niosh/docs/2002-149/pdfs/2002-149.pdf

Centers for Disease Control and Prevention. (2019). Antibiotic Resistance Threats in the United States, 2019. Atlanta, GA: U.S. Department of Health and Human Services, CDC. https://www.cdc.gov/drugresistance/pdf/threats-report/2019-ar-threats-report-508.pdf

Centers for Disease Control and Prevention. (2020a). Coronavirus (COVID-19). https://www.cdc.gov/coronavirus/2019-ncov/index.html

Centers for Disease Control and Prevention. (2020b). Reports of selected campylobacter outbreak investigations. https://www.cdc.gov/campylobacter/outbreaks/outbreaks.html

Centers for Disease Control and Prevention. (2020c). Leptospirosis. https://www.cdc.gov/leptospirosis/index.html

Centers for Disease Control and Prevention. (2020d). Guidance for reducing health risks to workers handling human waste or sewage https://www.cdc.gov/healthywater/global/sanitation/workers_handlingwaste.html

Centers for Disease Control and Prevention. (2020e). Immunization schedules. https://www.cdc.gov/vaccines/schedules/hcp/index.html

Clark, C. S. (1987). Potential and actual biological related health risks of wastewater industry employment. *Journal of Water Pollution Control Federation, 59*(12), 999.

Clark, C. S., Bjornson, H. S., Schwartz-Fulton, J., Holland, J. W., & Gartside, P. S. (1984). Biological health risks associated with the composting of wastewater treatment plant sludge. *Journal of Water Pollution Control Federation, 56*(12), 1269.

Cromeans, T. L., Kahler, A. M., & Hill, V. R. (2010). Inactivation of adenoviruses, enteroviruses, and murine norovirus in water by free chlorine and monochloramine. *Applied and Environmental Microbiology, 76*, 1028–1033.

Eftim, S. E., Hong, T., Soller, J., Boehm, A., Warren, I., Ichida, A., & Nappier, S. P. (2017). Occurrence of noroviruses in raw sewage: A systematic literature review and meta-analysis. *Water Research, 11*, 366–374. https:// doi.org/ 10.1128/AEM.01342-09

Environmental Leverage, Inc. (2003). Higher life forms or indicator organisms. https://environmentalleverage.com/Higher%20life%20forms.htm

Gerba, C. P., Tamimi, A. H., Pettigrew, C., Weisbrod, A. V., & Rajagopalan, V. (2011). Sources of microbial pathogens in municipal solid waste landfills in the United States of America. *Waste Management & Research, 29*(8), 781–790. https://doi.org/10.1177/0734242X10397968

Kiu, R., & Hall, L. J. (2018). An update on the human and animal enteric pathogen. *Emerging Microbes & Infections, 7*(1), 141 https://doi.org/10 .1038/s41426-018-0144-8

LeChevallier, M. W., Mansfield, T. J., & MacDonald Gibson, J. (2019). Protecting wastewater workers from disease risks: Personal protective equipment guidelines. *Water Environment Research, 92*(4), 1–10. https://doi.org/ 10.1002/wer.1249

Lin, K., & Marr, L. C. (2017). Aerosolization of Ebola virus surrogates in wastewater systems. *Environmental Science & Technology, 51*(5), 2669–2675.

Metcalf and Eddy (Eds.). (2007). *Water reuse: Issues, technologies, and applications*. New York, NY: McGraw-Hill.

Metcalf and Eddy, Inc. (2013). Wastewater engineering: Treatment and resource recovery (5th ed.). New York, NY: McGraw-Hill.

Niazi, S., Hassanvand, M. S., Mahvi, A. H., Nabizadeh, R., Alimohammadi, M., Nabavi, S., Faridi, S., Dehghani, A., Hoseini, M., Moradi-Joo, M., Mokamel, A., Kashani, H., Yarali, N., & Yunesian, M. (2015). Assessment of bioaerosol contamination (bacteria and fungi) in the largest urban wastewater treatment plant in the Middle East. *Environmental Science and Pollution Research, 22*(20), 16014–16021. https://doi.org/10.1007/ s11356-015-4793-z

Occupational Safety and Health Administration. (2002). Job Hazard Analysis. 29 C.F.R. 1910.132 (d),(1); https://www.osha.gov/Publications/osha 3071.pdf

Oppliger, A., Hilfiker, S., & Vu Duc, T. (2005). Influence of seasons and sampling strategy on assessment of bioaerosols in sewage treatment plants in Switzerland. *The Annals of Occupational Hygiene, 49*(5), 393–400. https:// doi.org/10.1093/annhyg/meh108

Pascual, L., Pérez-Luz, S., Yáñez, A., Santamaría, A., Gibert, K., Salgot, M., Apraiz, D., & Catalán, V. (2003). Bioaerosol emission from wastewater treatment plants. *Aerobiologia, 19*(3–4), 261–270. https://doi.org/10.1023/B:AERO.0000006598.45757.7f

Soda, E. A., Barskey, A. E., Shah, P. P., Schrag, S., Whitney, C. G., Arduino, M. J., Reddy, S. C., Kunz, J. M., Hunter, C. M., Raphael, B. H., & Cooley, L. A. (2017). Health Care–Associated Legionnaires' Disease Surveillance Data from 20 States and a Large Metropolitan Area—United States, 2015. *Morbidity and Mortality Weekly Report, 66*(22), 584–589.

Soller, J. A., Eftim, S. E., & Nappier, S. P. (2018). Direct potable reuse microbial risk assessment methodology: Sensitivity analysis and application to State log credit allocations. *Water Research, 128*, 286–292. https://doi.org/10.1016/j.watres.2017.10.034

Sun, J., Zhu, A., Li, H., Zheng, K., Zhuang, Z., Chen, Z., Shi, Y., Zhang, Z., Chen, S., Liu, X., Dai, J., Li, X., Huang, S., Huang, X., Luo, L., Wen, L., Zhuo, J., Li, Y., Wang, Y., Zhang, L., . . . Li, Y.-M. (2020) Isolation of infectious SARS-CoV-2 from urine of a COVID-19 patient. *Emerging Microbes & Infections, 9*(1), 991–993, DOI: 10.1080/22221751.2020.1760144

Thorn, J., & Kerekes, E. (2001). Health effects among employees in sewage treatment plants: A literature survey. *American Journal of Industrial Medicine, 40*(2), 170–179. https://doi.org/10.1002/ajim.1085

Tortora, G. J., Funke, B. R., & Case, C. L. (Eds.). (1982). *Microbiology: An Introduction.* Menlo Park, California: Benjamin/Cummings Publishing.

U.S. Environmental Protection Agency. (1991). Guidance manual for compliance with the filtration and disinfection requirements for public water systems using surface water sources. Office of Drinking Water, Washington, D.C.

U.S. Environmental Protection Agency. (2020a). Coronavirus and drinking water and wastewater. https://www.epa.gov/coronavirus/coronavirus-and-drinking-water-and-wastewater

U.S. Environmental Protection Agency. (2020b). Biosolids. https://www.epa.gov/biosolids

Washington On-site Sewage Association. (2020). Pathogens 2020 webinar on research grant on risks of exposure and best practices. https://www.doh.wa.gov/CommunityandEnvironment/WastewaterManagement/OnsiteSewageSystemsOSS

Water Environment Federation. (2001). *Wastewater Biology: The Microlife (2nd ed.).* Alexandria, Virginia: Water Environment Federation.

World Health Organization. (2020). Advice on the use of masks in the context of COVID-19. https://www.who.int/publications/i/item/advice-on-the-use-of-masks-in-the-community-during-home-care-and-in-healthcare-settings-in-the-context-of-the-novel-coronavirus-(2019-ncov)-outbreak

8.0   SUGGESTED READINGS

National Institutes of Health. (n.d.). Home Page. https://www.nih.gov/

National Institute for Occupational Safety and Health. (n.d.). Home Page. https://www.cdc.gov/niosh/index.htm

Occupational Safety and Health Administration General Duty Clause. (n.d.). https://www.osha.gov/laws-regs/oshact/section5-duties

Water Environment Federation. (2017). *Operation of Water Resource Recovery Facilities* (7th ed.). Manual of Practice No. 11. New York: McGraw-Hill.

2

Key Safety Considerations for COVID-19 and Other Biohazards

The considerations for addressing biohazards and PPE for wastewater personnel provided here are generally applicable to all biohazards but can be tailored to the specific risks presented by the novel human coronavirus that causes COVID-19 (SARS-CoV-2).

1.0 BACKGROUND ON ASSESSING RISKS IN THE WORKPLACE

The level of risk associated with any work task depends on the type and source of hazard, the extent of exposure, magnitude and frequency of the hazard or exposure, and the ability to minimize these exposures and the possible risks of adverse outcomes through informed decisions and actions. Resources for making informed decisions include international, national, state, and local guidance; regulatory requirements or recommendations; as well as relevant requirements, policies, and procedures of your organization. These resources may have relevant information for decision-making and actions, but it is ultimately up to your organization to specify and implement worker safety practices and procedures that are fit-for-purpose and sufficiently protective of all workers. Wastewater treatment personnel have a key role to play in protecting their health on the job and in the community.

Within wastewater systems, biological hazards to workers can be encountered through contact transfer (e.g., exposure to wastewater, biosolids or contaminated surfaces, especially of hands, mouth, nose, and eyes), splashes, whole body contact (e.g., accidental immersion), abrasions or cuts, aerosols, and dust. Because the specific biological hazards vary by job activity and wastewater system, it is necessary for facility managers and staff to conduct a Job Safety (or Hazard) Analysis (JSA/JHA) (Occupational Safety and Hazard Administration [OSHA], 2002). This analysis identifies each task in a job, defines the potential hazards, and outlines critical safety practices. Potential hazards can include physical, chemical, biological, electrical, and radiological sources, and gas/emissions. Each task-related hazard is ranked by probability, severity, and potential consequence. Once the hazards are known and prioritized, appropriate hazard control measures can be identified. As new hazards such as the SARS-CoV-2 virus emerge, existing control measures must be re-evaluated to determine if worker protections are adequate or must be changed to address both the existing and new hazard. Figure 2.1 provides the conceptual process for the JSA/JHA, as well as examples of exposure routes and hazard sources.

Figure 2.2 presents the hierarchy of hazard control measures, which can include engineering controls, administrative controls, training, required personal protective equipment (PPE), permits, and other health-related measures, such as vaccines and physical requirements. In the updated text of *Safety,*

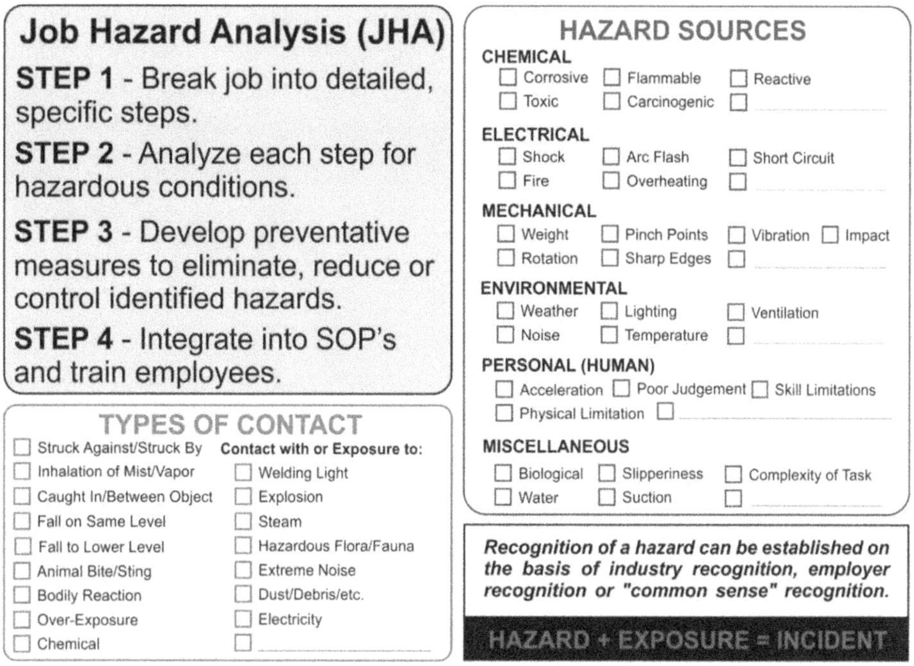

Job Hazard Analysis (JHA)

STEP 1 - Break job into detailed, specific steps.

STEP 2 - Analyze each step for hazardous conditions.

STEP 3 - Develop preventative measures to eliminate, reduce or control identified hazards.

STEP 4 - Integrate into SOP's and train employees.

TYPES OF CONTACT

☐ Struck Against/Struck By   **Contact with or Exposure to:**
☐ Inhalation of Mist/Vapor   ☐ Welding Light
☐ Caught In/Between Object   ☐ Explosion
☐ Fall on Same Level   ☐ Steam
☐ Fall to Lower Level   ☐ Hazardous Flora/Fauna
☐ Animal Bite/Sting   ☐ Extreme Noise
☐ Bodily Reaction   ☐ Dust/Debris/etc.
☐ Over-Exposure   ☐ Electricity
☐ Chemical   ☐ _____

HAZARD SOURCES

CHEMICAL
☐ Corrosive   ☐ Flammable   ☐ Reactive
☐ Toxic   ☐ Carcinogenic   ☐ _____

ELECTRICAL
☐ Shock   ☐ Arc Flash   ☐ Short Circuit
☐ Fire   ☐ Overheating   ☐ _____

MECHANICAL
☐ Weight   ☐ Pinch Points   ☐ Vibration   ☐ Impact
☐ Rotation   ☐ Sharp Edges   ☐ _____

ENVIRONMENTAL
☐ Weather   ☐ Lighting   ☐ Ventilation
☐ Noise   ☐ Temperature   ☐ _____

PERSONAL (HUMAN)
☐ Acceleration   ☐ Poor Judgement   ☐ Skill Limitations
☐ Physical Limitation   ☐ _____

MISCELLANEOUS
☐ Biological   ☐ Slipperiness   ☐ Complexity of Task
☐ Water   ☐ Suction   ☐ _____

Recognition of a hazard can be established on the basis of industry recognition, employer recognition or "common sense" recognition.

HAZARD + EXPOSURE = INCIDENT

FIGURE 2.1   Conceptual process for the Job Safety Analysis (Reprinted with permission by John Bannen).

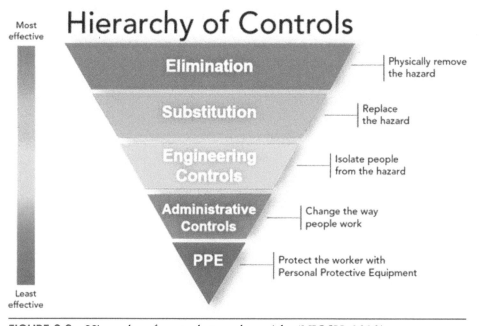

Hierarchy of Controls

Most effective

Elimination — Physically remove the hazard

Substitution — Replace the hazard

Engineering Controls — Isolate people from the hazard

Administrative Controls — Change the way people work

PPE — Protect the worker with Personal Protective Equipment

Least effective

FIGURE 2.2   Hierarchy of controls to reduce risks (NIOSH, 2020).

Health, and Security in Wastewater Systems, or MOP 1 (WEF, 2012) featured in this publication, specific guidance can be found in two key tables: Table 1.8 lists recommended PPE based on the various exposure routes, and Table 1.9 describes exposure routes for specific job activities. These tables may help identify appropriate task-specific PPE and other necessary control measures identified by JSAs.

2.0   GENERAL INFORMATION FOR WASTEWATER WORKERS

2.1   Should I Be Worried About Catching COVID-19 from Working with Wastewater or Biosolids?

 Information about the SARS-CoV-2 virus and the health risks it poses is currently limited but growing rapidly. Based on the available data on this virus and knowledge about similar viruses, experts from the World Health Organization (WHO), the Water Environment Federation (WEF), and U.S. Centers for Disease Control and Prevention (CDC), among others, agree that the occupational risk of infection to wastewater workers from the COVID-19 virus are low, and not greater than those from other pathogens typically present in wastewater. The heightened concern about wastewater safety has highlighted the importance of existing recommended precautions to prevent and control exposures to all wastewater personnel, and it serves as a reminder to review and (if needed) strengthen safety protocols and good hygiene practices. Scientists are still learning about this virus, and as new information becomes available, WEF will review these recommendations and updates will be published as appropriate and as needed.

2.2   What Are the Usual Precautions Against Biohazards?

   Default PPE includes waterproof gloves, waterproof boots, a uniform or waterproof coveralls, a covering for the mouth and nose (medical grade mask or respirator), and eye protection (e.g., safety glasses, goggles, or a face shield). With proper fitting, training, and maintenance, this PPE is intended to reduce risks from exposure to biohazards in wastewater and biosolids, treated or untreated, from collection systems to treatment and disposal.

2.3   How Do I Know If I Need Additional Precautions or Do Not Need as Many?

Risks are site- and job-specific: the type and level of hazards vary by task and by the conditions, equipment, and configuration at each utility. Because no two wastewater systems are the same, each utility should conduct a job safety analysis (JSA) to identify hazards and appropriate controls for each task. A JSA, also called a job hazard analysis (JHA), is a procedure to integrate accepted safety and health principles and practices into a particular task or job operation. For each basic step of the job, potential hazards are identified, and the safest ways are recommended to do the task.

2.4   Do I Need a Respirator?

Medical or dust masks can protect workers from splashes, sprays, and particulate materials (e.g., dust), and they can reduce the chances that sick workers will spread their illnesses. National Institute for Occupational Safety and Health (NIOSH)-approved filtering facepiece respirators, such as an N95 or similar, provide a higher level of protection; for example, they protect wearers from inhalation of aerosols. A JSA should be used to determine conditions where a respirator might be appropriate. It is important to remember that respirators have strict requirements for fitting and training, and a respirator used incorrectly may not provide better protection from aerosols than medical masks. Regardless of the type of mouth and nose protector, all such equipment must be cleaned and maintained in working order or replaced.

3.0   FREQUENTLY ASKED QUESTIONS ABOUT COVID-19 RISKS, WASTEWATER BIOHAZARDS, AND GENERAL WASTEWATER WORKER PRECAUTIONS

This section provides information on exposure routes and sources for the SARS-CoV-2 virus to help utilities assess the likely magnitude and probability of exposure of wastewater workers to this pathogen. It also covers general recommendations for worker safety.

3.1 Why Do Experts Believe That the Occupational Risk to Wastewater Workers from a Pathogen like the COVID-19 Virus (Also Called SARS-CoV-2) Is Relatively Low?

Based on the information available, experts from WHO, WEF, CDC, and other authoritative sources agree that infection risk to wastewater workers from the SARS-CoV-2 virus is low and not greater than risks from other pathogens typically present in wastewater. Taking the precautions that protect workers against typical wastewater pathogens should also adequately protect workers against the SARS-CoV-2 virus (WEF, 2020a; WEF, 2020b).

- The route(s) of transmission and exposure. Risk to wastewater workers increases if the pathogen is present in feces, remains infectious in feces and wastewater, and if it can be transmitted via aerosols because some wastewater-related activities generate aerosols. More information on these topics is provided below (WEF, 2020a; WEF, 2020b).

- Survival in the environment and through wastewater treatment processes. Pathogens that persist in wastewater or through treatment pose higher risks. More information on these topics is provided below. (WEF, 2020a; WEF, 2020b).

- Dose-response (likelihood of infection based on numbers of pathogens to which you are exposed). The risk of becoming infected increases as a person is exposed to greater numbers of a pathogen. Data specific to the COVID-19 virus are not yet available, so it is unknown how much the risk of infection increases as exposure to the COVID-19 viruses increases.

- Precautions. Precautions to reduce risk from pathogens include engineering and administrative controls and PPE. More information on PPE is provided below (WEF, 2020a; WEF, 2020b).

3.2 What Are the Main Routes of Transmission for the SARS-CoV-2 Virus?

The SARS-CoV-2 virus is mainly transmitted through respiratory droplets produced when an infected person coughs, sneezes, or talks, or by direct contact with an infected person, with possible further transmission by contact with contaminated surfaces (e.g., from deposited respiratory droplets). Transmission via wastewater aerosols, fecal wastes, wastewater, sludge, or biosolids is unlikely (WEF, 2020a; WEF, 2020b).

- The SARS-CoV-2 virus causes infection mainly from exposure to respiratory droplets that can be produced when an infected person coughs, sneezes, talks, sings, or exhales. These droplets can land in the mouth, nose, or eyes of others nearby or may be inhaled.

- It may be possible for a person to become infected by touching a surface or object that has the virus on it and then touching their own mouth, nose, or possibly their eyes.

- Transmission may also occur through direct contact with people who are sick. For example, a healthy person could become infected if they shake hands with someone whose hands are contaminated with respiratory secretions, and then touch their own nose, eyes, or mouth.

- The rate of transmission is quite high, and perhaps higher than common influenza (flu) (Sanche et al., 2020; WHO, 2020).

- Although the genetic material (RNA)of the SARS-CoV-2 virus has been found in the feces of sick patients, infectious virus has rarely been detected, and transmission via feces does not appear to be a significant route, based on both virological and epidemiological evidence. Respiratory secretions, urine, vomit, and eye secretions also may be sources of this virus, if present in wastewater and biosolids. To date, RNA from the SARS-CoV-2 virus has been detected in wastewater, but infectious or viable virus has not been (WEF, 2020a; WEF, 2020b).

3.3   Is the SARS-CoV-2 Virus Present in Wastewater, Sludge, and Biosolids, and Is It Infectious?

Infectious SARS-CoV-2 virus has not been detected in wastewater, sludge, and biosolids thus far; although the presence of infectious SARS-CoV-2 virus in these environments cannot be ruled out, it appears unlikely to be present at concentrations that would cause a measurable health risk (WEF, 2020a; WEF, 2020b).

- Infectious SARS-CoV-2 virus has been detected in human feces only rarely, and there is evidence that antiviral chemicals in the gut cause it to be rapidly inactivated (WEF, 2020a; WEF, 2020b; Zang et al., 2020). Hence, the presence of infectious COVID-19 virus in shed human feces is likely to be rare and in low concentrations.

- However, infectious SARS-CoV-2 virus is present in human respiratory fluids and secretions (saliva, mucus, etc.), which can also be shed and become a component of wastewater.

- The genetic material (RNA) of the SARS-CoV-2 virus has been detected in wastewater (WEF, 2020a; WEF, 2020b).

- However, infectious virus has not been detected in untreated wastewater, sludge, and biosolids so far, even though attempts are being made to detect it by many expert laboratories worldwide. While we cannot rule out the possibility that untreated wastewater, sludge, and biosolids (Class B) could contain some infectious SARS-CoV-2 virus, it appears unlikely to be present at concentrations that would cause a measurable health risk.

- While data are limited, there is no information to date that SARS-CoV-2 infections have resulted from contact with feces, wastewater, sludge, or biosolids.

3.4   How Long Does the SARS-CoV-2 Virus Survive in the Environment? Is It Expected to Survive in Wastewater Collection Systems? Does the SARS-CoV-2 Virus Present an Aerosol Exposure Risk to Wastewater Workers?

The risk of contracting COVID-19 from exposure to wastewater, sludge, and biosolids is considered low relative to other pathogens typically found in wastewater, as a result of the unlikely presence, expected dilution, and die-off of the virus in those matrices. The risk from aerosols in wastewater and biosolids systems is similarly low for these reasons (WEF, 2020a; WEF, 2020b).

- Data specific to survival of infectious SARS-CoV-2 virus in wastewater and biosolids environments, including collection systems, have not yet been collected. Like other coronaviruses and influenza viruses, this virus has an outer envelope (lipid membrane) that is vulnerable to damage that will make the virus unable to cause infection. Enveloped viruses are among the least environmentally stable of known viruses and other microorganisms and less likely to survive wastewater, sludge, biosolids, and water environments, as compared to other types of non-enveloped viruses.

- However, some related coronaviruses have been shown to survive for periods of time measurable in days to even weeks in wastewater, especially at lower temperatures (Wang et al., 2020; WEF, 2020a; WEF, 2020b).

- The survival of the SARS-CoV-2 virus has been studied in other aqueous media, on surfaces and in droplets and aerosols. Its survival on surfaces varies with the type of material, with complete (million-fold) loss of infectivity in 7 hours on paper, 1 day on wood and cloth, 2 days on glass and banknotes, and 7 days on stainless steel and plastic. Survival is similar to the SARS coronavirus (SARS-CoV-1) (Chin et al., 2020; van Doremalen et al., 2020; WEF, 2020a; WEF, 2020b).

- In aerosols generated in a laboratory, the virus survived for at least 3 hours in the air. However, in healthcare settings for COVID-19 virus patients and other patients with respiratory virus illnesses, evidence suggests that physical distancing of 1 to 2 m (3.3 to 6.6 ft) significantly reduces virus transmission from an infected patient (Canova et al., 2020; Chu et al., 2020).

3.5 Would the SARS-CoV-2 Virus Survive in Wastewater and Biosolids Treatment Systems If It Is Present?

Wastewater and biosolids treatment processes are expected to significantly remove and/or inactivate the SARS-CoV-2 virus, if it is present (WEF, 2020a; WEF, 2020b).

- Data specific to the SARS-CoV-2 virus have not yet been collected, but there is evidence for extensive removal and inactivation in wastewater and biosolids treatment processes of related viruses. The extent of virus reduction depends on the exact processes and how they are operated.

- Viruses are removed and inactivated by microbiological wastewater treatment processes such as activated sludge and membrane bioreactors. Some filtration processes also extensively remove viruses, including ultrafilters, nanofilters, and reverse osmosis membranes.

- Chemical disinfection processes, such as chlorination with free chlorine, disinfection with peracetic acid or chlorine dioxide, and ozonation can extensively inactivate viruses. Chloramines are a weaker disinfectant than free chlorine; the doses and contact times needed to effectively inactivate the SARS-CoV-2 virus using chloramines have not yet been established.

- Ultraviolet disinfection also extensively inactivates viruses, especially after secondary treatment processes.

- Viruses are inactivated in biosolids by thermal biological processes such as digestion and composting. Treatment requirements for Class A biosolids result in extensive virus reductions while treatments for Class B biosolids must produce extensive fecal bacteria reductions but the extent of virus reductions are not specified.

3.6   What Are the General Prevention Measures Against the COVID-19 Virus?

Follow the guidance from the Centers for Disease Control and Prevention (CDC, 2020b). Specifically,

- Wash your hands often with soap and water. If soap and water are not available, use a hand sanitizer with at least 60% alcohol.
- Always wash your hands after contact with respiratory secretions (for example, after blowing your nose, coughing, or sneezing) and at other critical times, such as after using the toilet or before preparing food.
- Avoid touching your eyes, nose, and mouth.
- Avoid gatherings of people outside your household, especially indoors.
- In general, stay at least 2 m (6 ft) away from others (social distancing).
- Wear a face covering in public.
- If you do not have a face covering, sneeze and cough into a bent elbow or tissue and then throw the tissue away and wash your hands.
- Clean and disinfect frequently touched and other potentially contaminated surfaces.
- Stay home if you are sick; do not infect co-workers!

3.7   What Are the Precautions and Safety Practices for Workers at Wastewater and Biosolids Facilities?

- Follow and practice the general preventive measures described in section 2.0 of this chapter and established by your utility.
- Also follow existing guidance, recommendations and related health and safety advice from sources such as WEF, OSHA, U.S. CDC, WHO, and relevant state and local sources.
- If there are site-specific factors or questions about the protections needed for any task, a job safety analysis (JSA) (also sometimes called a job hazard analysis) should be conducted to assess the risks and provide recommendations on the appropriate precautions. See JSA details below.
- Precautions for biohazards from wastewater and biosolids begin with engineering controls, followed by administrative controls, then PPE such as impermeable and waterproof gloves and boots, a medical grade face mask and face shield, goggles, and waterproof or other protective outer wear.

- Specifications for medical grade face masks are available from the American Society for Testing and Materials (ASTM International, 2017; 2019; 2020).
- NIOSH-approved filtering facepieces respirators, such as N95 respirators, are likely not needed for most activities; a job safety analysis can be conducted to identify any tasks that would require this level of respiratory protection.

4.0   CONSIDERATIONS AND QUESTIONS FOR WASTEWATER JOB SAFETY ANALYSIS

4.1   Key Questions to Identify and Evaluate Workplace, Job and Task-Specific Worker Hazards

1. Does your organization have a formal process and stepwise system to evaluate hazards in the workplace?

   a. If YES, follow those established procedures to identify the hazard.

   b. If NO, Refer to Chapter 3 of MOP 1 (WEF, 2012), "Identifying and Predicting Hazards".

2. What is the "exposure risk" of workers to task and exposure-specific hazards? Specifically:

   a. Where is the worker at risk, how is the worker at risk, and when is the worker at risk?

   b. What is the source, type, magnitude, frequency, duration, and probability of exposure?

   Such an analysis determines the severity of exposure and will aid in determining the appropriate protective methods to use to reduce the hazard. For example, pathogen exposure risks are greatest from untreated wastewater, sludge, and Class B biosolids, which have higher pathogen concentrations compared to Class A biosolids.

3. In determining the exposure risk, did you consider the following three elements?

   a. Workers performing the task,

   b. Determining why and how the task being performed increases the risk of exposure to a hazard, and

c. The environment in which the task is being performed and the hazard source.

4. If PPE was selected for worker protection, was a documented hazard assessment completed as required by OSHA?

   a. If NO, do a Job Hazard Analysis (also called Job Safety Analysis [JSA]) according to OSHA specifications.

   b. If YES, make sure the JSA is up to date and that it addresses any new or emerging hazard that has become apparent since the previous JSA. The JSA may need to be reviewed and possibly updated to address the new hazard.

4.2   Questions to Address in Developing Policies and Practices to Protect Workers

1. Does your utility have a safety plan, system, and manual or guide of practice for worker protection from all work hazards?

2. Are all work hazards addressed through administrative, engineering, and personal protective equipment (PPE) elements?

3. Is a Job Safety Analysis (also called a Job Hazard Analysis) done for all jobs and tasks, types of workers and environmental exposure sources and conditions?

   a. JSAs should be done. The procedures for them are established and available. See examples below.

4. If PPE was selected as an appropriate method for worker protection, were engineering controls and administrative controls evaluated beforehand on the basis of their relevance and effectiveness for selection and implementation of appropriate PPE?

5. For worker hazards related to exposure to wastewater, sludge, or biosolids (or environmental materials or media contaminated with them [e.g., primarily air and surfaces]), does PPE include the following key recommended items: waterproof boots (or similar footwear), waterproof gloves, a mask or respirator for mouth and nose, a face shield and goggles, and waterproof or water-repellant outerwear (e.g., coveralls, gown or apron)?

6. If changing the type of PPE or frequency of use is determined to be the appropriate course of action to protect the worker, did you account for the impact to worker efficiency and safety in completing the assigned task?

7. What training will the employees need on the new hazard protection?
   a. Equipment use, care, and maintenance
   b. Competency on procedural changes

8. If a respirator was identified as a needed worker PPE, does your organization currently have a written respiratory protection program that addresses work site specific procedures?

9. Does the new hazard protection implemented have any documentation or recordkeeping requirements? For example, employees wearing respirators must be medically evaluated and approved to wear a respirator (OSHA, 29 C.F.R. 1910.134) and must be fit tested on all the respirators worn in the workplace, both initially and annually thereafter. Events that may affect face shape, such as weight change or dental work, also require fit testing. Is there a system for respirator care, cleaning, and replacement, including record keeping?

It is important to note that some of the references and suggested readings provide information about general and overall workplace hazards from and exposure sources to SARS-CoV-2 virus and the COVID-19 disease as useful guidance from authoritative sources. However, they DO NOT necessarily provide sufficiently specific descriptions to address all possible sources of exposure to this virus and its disease for all possible work tasks and site-specific conditions that may be relevant to your job, your tasks, your specific utility, or your individual health/wellness and susceptibility status. Therefore, all PPE decisions should be made and implemented locally for each relevant job task and worker, tailored to utility conditions, resources, capacities, and local regulations.

Some of the references and suggested readings serve as general resources, but do not provide task-specific guidance on PPE. It is the responsibility of the utility to develop JSAs/JHAs that define the PPE requirements for their workers, based on site-specific and activity-specific conditions.

5.0   REFERENCES

ASTM International. (2017). Designation: F2100 – 19. Standard Specification for Performance of Materials Used in Medical Face Masks. https://www.astm.org/Standards/F2100.htm

ASTM International. (2019). Designation: F2101 – 19. Standard Test Method for Evaluating the Bacterial Filtration Efficiency (BFE) of Medical Face Mask Materials, Using a Biological Aerosol of Staphylococcus aureus. https://www.astm.org/Standards/F2101.htm

ASTM International. (2020). Designation: F2100 – 19. Standard Specification for Performance of Materials Used in Medical Face Masks. https://www.astm.org/Standards/F2100.htm

Canova, V., Schläpfer, L. H., Piso, R. J., Droll, A., Fenner L., Hoffmann, T., & Hoffmann, M. (2020). Transmission risk of SARS-CoV-2 to healthcare workers—Observational results of a primary care hospital contact tracing. *Swiss Medical Weekly*. 2020;150:w20257. https://doi.org/10.4414/smw.2020.20257

Centers for Disease Control and Prevention. (2015). Guidance for reducing health risks to workers handling human waste or sewage. https://www.cdc.gov/healthywater/global/sanitation/workers_handlingwaste.html

Centers for Disease Control and Prevention. (2020a). Information for sanitation and wastewater workers on COVID-19. https://www.cdc.gov/coronavirus/2019-ncov/community/sanitation-wastewater-workers.html

Centers for Disease Control and Prevention. (2020b). Coronavirus Disease: Protect Yourself. https://www.cdc.gov/coronavirus/2019-ncov/prevent-getting-sick/prevention.html

Chin, A. W. H., Chu, J. T. S., Perera, M. R.A., Hui, K. P. Y., Yen, H.-L., Chan, M. C. W., Peiris, M., & Poon, Leo L. M. (2020). Stability of SARS-CoV-2 in different environmental conditions. *The Lancet Microbe, 1*(1). https://doi.org/10.1016/S2666-5247(20)30003-3

Chu, D. K., Akl, E.A., Duda, S., Solo, K., Yaacoub, S., & Shünemann, H. J. (2020). Physical distancing, face masks, and eye protection to prevent person-to-person transmission of SARS-CoV-2 and COVID-19: A systematic review and meta-analysis. *Lancet*. 2020; (published online June 1.) https://doi.org/10.v1016/S0140-6736(20)31142-9

National Institute of Occupational Safety and Health. (2020). Hierarchy of Controls. https://www.cdc.gov/niosh/topics/hierarchy/default.html

Occupational Safety and Health Agency. (2002). Job Hazard Analysis (OSHA 3071). https://www.osha.gov/Publications/osha3071.pdf

Occupational Safety and Health Agency. (2020). Guidance on preparing workplaces for COVID-19 (OSHA 3990-03). https://www.osha.gov/Publications/OSHA3990.pdf

Sanche S., Lin, Y. T., Xu, C., Romero-Severson, E., Hengartner, N., & Ke, R. (2020). High contagiousness and rapid spread of severe acute respiratory syndrome coronavirus 2. *Emerging Infectious Diseases, 26*(7), 1470–1477. https://dx.doi.org/10.3201/eid2607.200282

van Doremalen, N., Bushmaker, T., Morris, D. H., Holbrook, M. Gamble, A., Williamson, B. N., Tamin, A., Harcourt, J. L., Thornburg, N. J., Gerber, S. J., Lloyd-Smith, J.O., & de Witt, E. (2020). Aerosol and surface stability of SARS-CoV-2 as compared with SARS-CoV-1. *New England Journal of Medicine, 382*(16), 1564–1567. https://www.nejm.org/doi/full/10.1056/NEJMc2004973

Wang, X.-W., Li, J.-S., Guo, T.-K., Zhen, B., Kong, Q.-X., Yi, B., Li, Z., Song, N., Jin, M., Wu, X.-M., Xiao, W.-J., Zhu, X.-M., Gu, C.-Q., Yin, J., Wei, W., Yao, W., Liu, C., Li, J.-F., Ou, G.-R. . . . Li, J.-W. (2005). Excretion and detection of SARS coronavirus and its nucleic acid from digestive system. *World Journal of Gastroenterology, 11*(28), 4390–4395. https://www.wjgnet.com/1007-9327/full/v11/i28/4390.htm

Water Environment Federation. (2012). *Safety, health, and security in wastewater systems* (6th ed., Manual of Practice No. 1). Alexandria, Viriginia: Water Environment Federation.

Water Environment Federation. (2020a). Coronavirus and Water Systems. An update and expansion on "The Water Professional's Guide to COVID-19". https://www.wef.org/news-hub/wef-news/coronavirus-and-water-systems

Water Environment Federation. (2020b). Residuals and biosolids issues concerning COVID-19 Virus. https://www.wef.org/news-hub/wef-news/residuals-and-biosolids-issues-concerning-covid-19-virus/

World Health Organization. (2020). Q&A: Influenza and COVID-19—similarities and differences. https://www.who.int/emergencies/diseases/novel-coronavirus-2019/question-and-answers-hub/q-a-detail/q-a-similarities-and-differences-covid-19-and-influenza

Zang, R., Castro, F. G., McCune, B. T., Zeng, Q., Rothlauf, P. W., Sonnek, N. M., Liu, Z., Brulois, K. F., Wang, X., Greenberg, H. B., Diamond, M. S., Ciorba, M. A., Whelan, S. P. J., & Ding, S. (2020). TMPRSS2 and TMPRSS4 mediate SARS-CoV-2 infection of human small intestinal enterocytes. *Science Immunology, 5.* https://immunology.sciencemag.org/content/immunology/5/47/eabc3582.full.pdf

6.0   SUGGESTED READINGS

LeChevallier, M. W., Mansfield, T. J., & MacDonald Gibson, J. (2019). Protecting wastewater workers from disease risks: Personal protective equipment guidelines. *Water Environment Research, 92*(4), 1–10. https://doi.org/10.1002/wer.1249

Li, Y., Wong, T., Chung, J., Guo, Y. P., Hu, J. Y., Guan, Y. T., Yao, L., Song, Q. W., & Newton, E. (2006). In vivo protective performance of N95 respirator and surgical facemask. *American Journal of Industrial Medicine, 49*(12), 1056–1065. https://doi.org/10.1002/ajim.20395

World Health Organization. (2020a). Water, sanitation, hygiene, and waste management for the COVID-19 virus: Interim guidance, 23 April 2020. COVID-19: Infection prevention and control / WASH. https://www.who.int/publications/i/item/water-sanitation-hygiene-and-waste-management-for-the-covid-19-virus-interim-guidance

World Health Organization. (2020b). Advice on the use of masks in the context of COVID-19 Interim guidance. 5 June 2020. COVID-19: Infection prevention and control / WASH. https://www.who.int/publications/i/item/advice-on-the-use-of-masks-the-community-during-home-care-and-in-health-care-settings-in-the-context-of-the-novel-coronavirus-(2019-ncov)-outbreak

3

Research Needs for COVID-19

1.0 WHAT WE KNOW THAT CAN BE REDUCED TO PRACTICE

Chapters 1 and 2 focus on current knowledge that can be reduced to practice. The focus of this chapter is on research needs that, once filled, can modify or further improve practice.

2.0 WHAT WE KNOW THAT WE DO NOT KNOW

In this section, we will summarize facets where our knowledge needs to be improved concerning biological hazards in wastewater.

2.1   Entire Water Cycle Approach

Infections caused by human coronaviruses are typically transmitted through respiratory droplets (Dwosh et al., 2003). However, as a result of the longevity of coronaviruses on numerous media such as surfaces, moderately acidic aqueous matrices such as bodily fluids, and feces (Geller et al., 2012), coronavirus-mediated respiratory infections can also be transmitted by exposure to such media. In addition, aerosolized human coronavirus survival is enhanced under low-moisture, high-temperature environments (Ijaz et al., 1985). Human coronaviruses can retain their infection potential for extended durations in aqueous media and in wastewater streams as well, thereby necessitating an investigation of wastewater streams and treatment processes as potential harbors or reservoirs that enhance human exposure (Casanova et al., 2009).

Of the different countermeasures against human coronavirus published in peer-reviewed literature, free chlorination and chloramination at high doses, ethanol-based treatment (Kariwa et al., 2006), and treatment with acetic acid (Rabenau et al., 2005) are shown to be effective. Bleach by itself was shown to be ineffective (Hulkower et al., 2011). In general, the efficacy of these treatments is compromised by the presence of organics or other "shielding" agents in the system (Rabenau et al., 2005). Such evidence of low vulnerability of human coronaviruses under conditions pertinent to engineered wastewater treatment systems (i.e., presence of colloids or particulate matter in the streams; different disinfection strategies practiced, including no disinfection) is concerning. Furthermore, the effect of treatment processes aimed at the liquid phase (activated sludge) and solids phase (anaerobic digestion) on human coronavirus survival and propagation are, as yet, poorly understood.

2.2   Risks to the General Public from Maintenance Activities in the Collection System (Such as Jetting and Aerosol Generation)

Both the utility worker and the general public should be aware that aerosols derived from raw wastewater can be present in areas where sewer maintenance activities occur. It is uncertain, however, whether infectious transmission of the SARS-CoV-2 virus that causes the COVID-19 disease occurs via these aerosols. To date, research specific to this question has not been conducted. Because of the uncertainty, it is recommended that collection system workers conducting maintenance activities adhere to the proper use of the appropriate personal protection equipment (PPE), including masks and gloves, and exercise best practices such as disinfecting vehicle interiors and removing clothing exposed to the aerosols during the activity before returning to other facilities or duties. Furthermore, the vicinity of the work area from which aerosols may be present should be clearly demarcated and adequately separated from general public traffic using markers, pylons, tapes, etc.

2.3 How Pathogens Partition from Bulk Liquid Phase into Aerosols (for Example, in Raw, Primary, and Secondary Wastewater) and onto Sludge

2.3.1 Partitioning of Viruses Between the Solid and Liquid Phases in Wastewater

From a size perspective, viruses are in the sub-colloidal range (Madigan & Martinko, 2006), and as such they could exist in both the liquid and solid phases in wastewater streams. However, considerable temporal variability exists even within the same water resource recovery facility (WRRF), at the same sampling location, in the relative viral fractions associated with the solid and liquid phases (Wellings et al., 1976). There is evidence that the partitioning of viruses amongst these two phases is affected by whether the viruses are enveloped or not. Specifically, it has been reported that enveloped viruses partition to a higher extent than non-enveloped viruses to the solids phase in wastewater samples (Ye et al., 2016). The same study also presented the kinetics of sorption and the extent of inactivation in this context (Ye et al., 2016). Additionally, the extent of partitioning itself varies as a function of the size distribution of the particles as well as the treatment operations themselves (Hejkal et al., 1981).

In a previous study, it was shown that most enteric viruses in wastewater streams and through treatment processes were present either free or in association with particles smaller than 0.3 µm in characteristic size (Hejkal et al., 1981). On the other hand, it has been shown that in effluent streams from secondary wastewater treatment processes, viruses were associated with particles larger than 8 µm or smaller than 0.65 µm characteristic size (Gerba et al., 1978).

The outer envelope of viruses has a composition similar to that of eukaryotic and prokaryotic membranes, including amphipathic phospholipids (Madigan & Martinko, 2006). The envelope also contains attachment proteins that help in colonizing the host cells (Madigan & Martinko, 2006). Specifically, in SARS-CoV-2, the spike glycoprotein is responsible for binding of receptors on host cells and entry of the virion into the host cell (Wrapp et al., 2020).

2.3.2 Implications of Partitioning

2.3.2.1 Shielding and protection from disinfectants

Viral association with particles may serve to protect the viruses from disinfectants either by shielding, in the case of ultraviolet (UV), or by hindering chemical penetration. These characteristics have been reviewed by Templeton, Andrews, and Hofmann (Templeton et al., 2008).

2.3.2.2 Transport in effluent streams (solid and liquid)

When associated with particles during treatment or in receiving waters, the transport of viruses will be governed by particle behavior (Brookes et al., 2004; Ferguson et al., 2003). Therefore, hydrodynamic models of transport, as well as treatment process models, must consider such associations.

2.3.2.3 Association of viruses with particles and exposure

It is likely that association of viruses with particles is non-specific relative to the host-receptor-specific binding, which involves structural, morphological, and chemical interactions (Dimitrov, 2004). A recent study has evaluated sorptive transfer of viruses between the skin and the aqueous phase for both enveloped and non-enveloped viruses (Pitol et al., 2017). In this study, liquid type, virus concentration, and wetness of skin were determinants in the transfer of viruses between the two phases (Pitol et al., 2017).

2.4 How to Interpret Molecular Signals in Various Media with Respect to Viability and Infectivity

All reports to date of SARS-CoV-2 in the water/wastewater environment have been based on molecular detections of the viral RNA. This is a result of the increased speed and decreased cost of molecular assays compared to culture-based assays. In addition, there have been limited reports of infectious SARS-CoV-2 in stool (Xiao et al., 2020), while reported RNA shedding rates in stool have been as high as 89% (Wölfel et al., 2020). While the presence of the viral genome is necessary for an infectious virus, it is not sufficient; for example, damage to a viral capsid would eliminate infectivity of the virus, but not the polymerase chain reaction (PCR) signal. Infectious virus is typically a fraction of molecular genome counts in environmental samples. There is no direct conversion between infectious virus and PCR detections. The correlation between molecular viral genome counts and infectious virus varies based on the virus and sampled matrix. Aging that may inactivate the virus but leaves untouched the remaining viral genome would increase the ratio between molecular genome counts and the infectious virus. As such, it is unlikely that a single ratio will be sufficient to consistently convert between molecular and infectious viral counts.

Methodologies exist to remove PCR detections of nucleic acids from damaged virions, potentially providing a pathway to connect molecular and infectious viral quantifications. Example methodologies include applying deoxyribonuclease (DNase)/ribonuclease (RNase) to digest free nucleic acids and applying intercalating dyes such as propidium monoazide to block PCR amplification of free nucleic acids. These methodologies have recently been reviewed (Emerson et al., 2017); however, the authors of this publication are not aware of prior work demonstrating the suitability of these methods for environmental detections of enveloped viruses.

Reports, primarily for surveillance purposes, have begun to emerge of SARS-CoV-2 RNA in wastewater streams and sludge (Ahmed, Angel, et al., 2020). Caution should be exercised not to extend these detections to infer viable virus. Critical work is necessary to determine if viable SARS-CoV-2 is present in wastewater and at which concentrations. Additional approaches to remove PCR detections of damaged SARS-CoV-2 virions may also help to unify these measurements. Optimization of concentration (Ahmed, Bertsch, et al., 2020) techniques as well as amplification and detection protocols are also needed. Because current state-of-the-science appears promising for collection system-wide surveillance, wastewater facilities could be an important contributor to these data sets by engaging in sampling of wastewaters following documented protocols.

2.5 SARS-CoV-2 Sensitivity to Disinfection or Stress in the Water Environment

No data are yet available on SARS-CoV-2 persistence or disinfection kinetics in the water environment. However, there has been prior work on survival of SARS-CoV-1 in aqueous systems (Casanova et al., 2009). It is critical that future experiments investigating SARS-CoV-2 persistence and disinfection assess viable virus. Given the enveloped structure of SARS-CoV-2, the expectation is that the environmental persistence and required disinfection CT values will be lower than those associated with non-enveloped viruses; however, it is critical to develop these primary data as an assurance that current practice is effective. Persistence should be assessed under a variety of conditions, including varying water quality and temperatures. Disinfection should be assessed similarly and include a variety of disinfectants. Subsets of these experiments are underway that should help to respond to the most immediate questions regarding SARS-CoV-2 in the water environment. Finally, we call on research funding agencies to fund this critical work.

While data on SARS-CoV-2 persistence and disinfection in water matrices is currently limited, we may be able to gain insights into likely viral fate using surface persistence data. A recent study investigating SARS-CoV-2 surface stability found no statistically significant difference from SARS-CoV-1 surface persistence (van Doremalen et al., 2020), suggesting that SARS-CoV-2 high infectivity is not due to enhanced resistance to environmental stressors. While we would expect this observation to extend to water matrices, additional research is necessary to validate these presumptions.

2.5.1 Chloramines Versus Free Chlorine

An important issue to consider here is the form of +1-valent chlorine that will dominate in a wastewater setting. In most circumstances, there will be sufficient ammonia-N present in treated effluents (even after nitrification) to allow effective conversion of free chlorine to monochloramine (NH2Cl). The inorganic chloramines, including NH2Cl, tend to be less effective disinfectants than free chlorine.

Disinfection of wastewater is practiced in the United States to ensure that surface waters are swimmable. Indicator bacterial standards, such as *Escherichia coli* and enterococcus, are used to ensure an acceptable level of swimming-associated rates of gastroenteritis. Respiratory illnesses, of unknown cause, are also associated with swimming in surface waters.

In general, coronaviruses are believed to be less or no more resistant to disinfectants than the non-enveloped enteric viruses having similar structure and morphology. However, in a recent review, Silverman and Boehm (2020) concluded that there is very little data available in the literature on the kinetics of coronavirus inactivation in water for disinfectants that are commonly used by the wastewater industry (Silverman & Boehm, 2020). They could not identify any experiments on coronavirus inactivation in water by chlorine or UV radiation. One study using wastewater found that SARS-CoV-1 was inactivated faster than *E. coli* or coliphage f2, but rate constants could not be developed from the data (Wang et al., 2005).

While it is believed that current wastewater disinfection processes are likely to be effective against coronaviruses, quantitative data for comparison to the more commonly studied enteric viruses is lacking and is needed to have the assurances that treatment is sufficient to control the risk from coronaviruses.

2.5.2 Needs Regarding Ultraviolet and Other Disinfectants

No data are yet available on SARS-CoV-2 to define UVC dose-response behavior (i.e., disinfection kinetics) of SARS-CoV-2 in aqueous suspension or on surfaces. However, data are available to describe the kinetics of inactivation of other related viruses. These previous reports suggest that SARS-CoV-2 should be effectively inactivated by UVC radiation at doses that are relevant to most disinfection applications.

Similarly, no data are available to define the action spectrum (i.e., wavelength-dependence) of SARS-CoV-2 for UVC radiation. This information will be needed to assess the effectiveness of UVC sources that have been developed as alternatives to conventional low-pressure (LP) mercury (Hg) lamps, which emit essentially monochromatic radiation at a characteristic wavelength of 254 nm. Data are needed to define UVC dose-response behavior at other wavelengths specifically to facilitate the use of common alternatives to LP Hg lamps, including UV LEDs and plasma (excimer) lamps.

No data appear to be available for inactivation in water by chlorine dioxide or ozone.

2.5.3 Fate in Reuse Systems

Water reuse systems are designed to achieve a high level of virus removal (historically targeting enteroviruses). The U.S. state of California has assigned default virus log reduction values of 6 for free chlorine disinfection, 6 for UV/AOP processes, and 1.5 for reverse osmosis (Trussell et al., 2019). Particularly for reverse osmosis, this may be highly conservative; however, information specifically on removal of enveloped viruses, such as SARS-CoV-2, in various treatment processes and trains used for reuse is not available.

2.6   Risk Assessment Approaches

Frameworks for quantitative microbial risk assessment (QMRA) in the context of the air exposure to pathogens in the wastewater environment have been developed. Haas et al. (2017) estimated the risk to workers in the collection system in vicinity to the inlet from a hospital treating Ebola patients. Noteworthy in the approach was the incorporation of a distribution for the ratio between gene copies and infectious virus, which is a similar problem to what would be faced with doing a QMRA for SARS-CoV-2. Carducci et al. (2018) did a QMRA (focusing on adenovirus) for wastewater treatment facility operators, and they determined that the greatest risks were at the influent and the aeration tank, with this dominant uncertainty arising from the concentration estimates for virus. Because SARS-CoV-2 has a lipid envelope, one unknown is whether this would enhance partitioning of the virus from the water phase to aerosols.

Stellacci et al. (2010) performed a QMRA (for virus, Campylobacter, and protozoa) for air exposure from wastewater treatment facilities to neighboring residents. This study illustrates how air dispersion models could be coupled with a source term to assess risks to the general community. Jahne et al. (2015) did a similar QMRA (focusing on bacterial pathogens) in the context of dispersion to neighbors from land application of manure.

Thus, frameworks for conducting QMRA for SARS-CoV-2 in the context of wastewater treatment plants exist; however, gaps in input data could be filled to gain more surety of the estimates.

2.7   Personal Protective Equipment

Currently, there is no evidence of viable SARS-COV-2 in untreated wastewater, treated wastewater, or sludges. However, viable virus has been found in feces of ill COVID-19 patients (Xiao et al., 2020). RNA from SARS-CoV-2 has been found in toilets from a hospital room in which a SARS-CoV-2 patient was in residence (Ong et al., 2020). A targeted program of sampling is needed to determine the strength of current opinion that viable virus is unlikely to be in materials currently managed by WRRFs—absence of evidence is not evidence. As it relates to PPE research, recent information has become available, and below are two key recommendations for research (LeChevallier et al., 2019).

2.8   Research Recommendations

2.8.1   Develop, Fund, and Conduct a Prospective Epidemiological Study of Infectious Disease Incidence Among Wastewater and Collection System Workers

Although wastewater treatment workers are exposed to higher levels of certain health outcomes relative to the general population, robust statistical analyses of infectious disease rates for U.S. wastewater workers are limited (LeChevallier et al., 2019). The study should at a minimum a) establish a baseline level of PPE use among wastewater and collection system workers, and b) examine observed health outcomes with respect to reported PPE use to test the efficacy of PPE as currently practiced in reducing infectious disease risks (LeChevallier et al., 2019).

2.8.2   Develop, Fund, and Conduct a Study to Characterize Respiratory Exposure for Typical Tasks Performed by Workers in Wastewater Collection and Treatment Operations

The presence and magnitude of infectious agents in tasks performed by wastewater workers is not well understood even though emerging research indicates that aerosolization of wastewater may expose them (Lee et al., 2016). Research is needed to "characterize the cumulative exposure of workers to aerosolized wastewater that may contain infectious agents" (Carducci et al., 2018).

3.0 WHAT WE DO NOT KNOW THAT WE DO NOT KNOW—RESEARCH RECOMMENDATIONS

3.1 Fate and Transport

There is little to no information on the nature of the interactions between enveloped viruses and non-host materials related to wastewater or biosolids matrices. As such, knowledge of mechanisms, kinetics, thermodynamics, and models of virus association as a function of particle size, surface characteristics, and chemistry will be beneficial.

3.2 Disinfection and Treatment

1. Ct (or It) data for the commonly used disinfectants should be developed for chlorine, chloramines, and UV light because these are the most commonly used disinfectants by the wastewater industry. (SARS-CoV-2 or closely related surrogate could be used, such as human coronavirus 229E). The effect of temperature and pH should be taken into consideration in the design of these experiments.

2. Dose-response data should also be developed for UV light with doses commonly used in wastewater treatment.

3. Lipid-containing viruses like coronaviruses may be more likely to clump and associate with suspended particulate organic matter than less hydrophobic non-lipid viruses. Thus, these should also be considered in the design of disinfectant studies.

4. Lipid-containing viruses are more likely to be attracted to the air-water interface, which may reduce the inactivation rate of these viruses in water. Should this be taken into consideration in the design of disinfectant studies? This also could be affected by the amount of organic matter present in the water.

3.3 Risk and Exposure Assessment

1. Size-resolved generation rates for airborne particles containing SARS-CoV-2 produced in various collection system operation and WRRF settings (e.g., bar screens, hydraulic jumps, cleaning operations, aeration chambers, etc.) need to be developed.

2. Dose-response information for the assessment of risk from SARS-CoV-2 needs to be developed.

3. Data on removal of SARS-CoV-2 by various types of PPE are needed to quantitatively assess degree of worker protection.

4. Estimation of the ratio between gene copies and infectious virus after various environmental transport conditions, and correlated, as appropriate, to the associated water matrix.

3.4   Preparedness

The wastewater industry is at the "end of the line" for inputs from a community undergoing an outbreak. While we do not know which "black swan" events may occur in the future, including outbreaks of other infectious diseases, it would be important to harvest lessons learned by the utility industry in response to the COVID-19 pandemic. This should be done formally to assess what might have worked to a greater or lesser degree and could serve as a roadmap for responses in the event of future stresses yet unknown.

4.0   REFERENCES

Ahmed, W., Angel, N., Edson, J., Bibby, K., Bivins, A., O'Brien, J. W., Choi, P. M., Kitajimae, M., Simpson, S. L., Li, J., Tscharke, B., Verhagen, R., Smith, W. J. M., Zaugg, J., Dierens, L., Hugenholtz, P., Thomas, K. V., & Mueller, J. F. (2020). First confirmed detection of SARS-CoV-2 in untreated wastewater in Australia: A proof of concept for the wastewater surveillance of COVID-19 in the community. *Science of The Total Environment*, *728*, 138764.

Ahmed, W., Bertsch, P., Bivins, A., Bibby, K., Farkas, K., Gathercole, A., Haramoto, E., Gyawali, P., Korajkic. A., McMinn, B. R., Mueller, J. F., Simpson, S. L., Smith, W., Symonds, E. M., Thomas, K. V., Verhagen, R., & Kitajimal, M. (2020). Comparison of virus concentration methods for the RT-qPCR-based recovery of murine hepatitis virus, a surrogate for SARS-CoV-2 from untreated wastewater. *Science of The Total Environment*, 139960.

Brookes, J. D., Antenucci, J., Hipsey, M., Burch, M. D., Ashbolt, N. J., & Ferguson, C. (2004). Fate and transport of pathogens in lakes and reservoirs. *Environment International*, *30*(5), 741–759.

Carducci, A., Donzelli, G., Cioni, L., Federigi, I., Lombardi, R., & Verani, M. (2018). Quantitative microbial risk assessment for workers exposed to bioaerosol in wastewater treatment plants aimed at the choice and setup of safety measures. *International Journal of Environmental Research and Public Health*, *15*(7), 1490.

Casanova, L., Rutala, W. A., Weber, D. J., & Sobsey, M. D. (2009). Survival of surrogate coronaviruses in water. *Water Research*, *43*, 1893–1898.

Dimitrov, D. S. (2004). Virus entry: Molecular mechanisms and biomedical applications. *Nature Reviews Microbiology*, *2*(2), 109–122.

Dwosh, H. A., Hong, H. H. L., Austgarden, D., Herman, S., & Schabas, R. (2003). Identification and containment of an outbreak of SARS in a community hospital. *CMAJ : Canadian Medical Association journal = journal de l'Association medicale canadienne*, *168*(11), 1415–1420.

Emerson, J. B., Adams, R. I., Román, C. M. B., Brooks, B., Coil, D. A., Dahlhausen, K., Ganz, H. H., . . . Rothschild, L. J. (2017). Schrödinger's microbes: Tools for distinguishing the living from the dead in microbial ecosystems. *Microbiome*, *5*(86), 1–23.

Ferguson, C., Husman, A. M. de R., Altavilla, N., Deere, D., & Ashbolt, N. (2003). Fate and transport of surface water pathogens in watersheds. *Critical Reviews in Environmental Science and Technology*, *33*(3), 299–361.

Geller, C., Varbanov, M., & Duval, R. (2012). Human coronaviruses: Insights into environmental resistance and its influence on the development of new antiseptic strategies. *Viruses*, *4*(11), 3044–3068.

Gerba, C. P., Stagg, C. H., & Abadie, M. G. (1978). Characterization of sewage solid-associated viruses and behavior in natural waters. *Water Research*, *12*, 805–812.

Haas, C. N., Rycroft, T., Bibby, K., & Casson, L. (2017) Risks from ebolavirus discharge from hospitals to sewer workers. *Water Environment Research*, *89*(4), 357–368. https://doi.org/10.2175/106143017X14839994523181

Hejkal, T. W., Wellings, F. M., Lewis, A. L., & LaRock, P. A. (1981). Distribution of viruses associated with particles in waste water. *Applied and Environmental Microbiology*, *41*(3), 628–634.

Hulkower, R. L., Casanova, L. M., Rutala, W. A., Weber, D. J., & Sobsey, M. D. (2011). Inactivation of surrogate coronaviruses on hard surfaces by health care germicides. *American Journal of Infection Control*, *39*(5), 401–407.

Ijaz, M. K., Brunner, A. H., Sattar, S. A., Nair, R. C., & Johnson-Lussenburg, C. M. (1985). Survival characteristics of airborne human coronavirus 229E. *Journal of General Virology*, *66*(2).

Jahne, M. A., Rogers, S. W., Holsen, T. M., Grimberg, S. J., & Ramler, I. P. (2015). Emission and dispersion of bioaerosols from dairy manure application sites: Human health risk assessment. *Environmental Science & Technology*, *49*(16), 9842–9849.

Kariwa, H., Fujii, N., & Takashima, I. (2006). Inactivation of SARS coronavirus by means of povidone-iodine, physical conditions and chemical reagents. *Dermatology, 212* (Suppl. 1), 119–123. doi: 10.1159/000089211

LeChevallier, M. W., Mansfield, T. J., & MacDonald Gibson, J. (2019). Protecting wastewater workers from disease risks: Personal protective equipment guidelines. *Water Environment Research, 92*(4). https://doi.org/10.1002/wer.1249

Lee, M. T., Pruden, A., & Marr, L.C. (2016). Partitioning of viruses in wastewater systems and potential for aerosolization. *Environmental Science & Technology Letters, 3*(5), 210–215. doi:10.1021/acs.estlett.6b00105

Madigan, M. T., & Martinko, J. M. (2006). *Brock biology of microorganisms* (15th edition). Prentice Hall.

Ong, S. W. X., Tan, Y. K., Chia, P. Y., Lee, T. H., Ng, O. T., Wong, M. S. Y., & Marimuthu, K. (2020). Air, surface environmental, and personal protective equipment contamination by severe acute respiratory syndrome coronavirus 2 (SARS-CoV-2) from a symptomatic patient. *JAMA, 323*(16), 1610–1612. doi:10.1001/jama.2020.3227

Pitol, A. K., Bischel, H. N., Kohn, T., & Julian, T. R. (2017). Virus transfer at the skin-liquid interface. *Environmental Science & Technology, 51,* 14417–14425.

Rabenau, H. F., Cinatl, J., Morgenstern, B., Bauer, G., Preiser, W., & Doerr, H. W. (2005). Stability and inactivation of SARS coronavirus. *Medical Microbiology and Immunology, 194*(1), 1–6.

Silverman, A. I., & Boehm, A. B. (2020). Systematic review and meta-analysis of the persistence and disinfection of human coronaviruses and their viral surrogates in water and wastewater. *Environmental Science & Technology Letters.* https://doi.org/10.1021/acs.estlett.0c00313

Stellacci, P., Liberti, L., Notarnicola, M., & Haas, C. N. (2010). Hygienic sustainability of site location of wastewater treatment plants: A case study. II. Estimating airborne biological hazard. *Desalination, 253*(1–3), 106–111.

Templeton, M. R., Andrews, R. C., & Hofmann, R. (2008). Particle-associated viruses in water: Impacts on disinfection processes. *Critical Reviews in Environmental Science and Technology, 38,* 137–164.

Trussell, R. S., Lai-Bluml, G., Chaudhuri, M., & Johnson, G. (2019). Developing a regional recycled water program in Southern California. *Water Practice and Technology, 14*(3), 570–578.

van Doremalen, N., Bushmaker, T., Morris, D. H., Holbrook, M. G., Gamble, A., Williamson, B. N., Tamin, A., Harcourt, J. L., Thornburg, N. J.,

Gerber, S. I., Lloyd-Smith, J. O, de Wit, E., & Munster, V. J. (2020). Aerosol and surface stability of SARS-CoV-2 as compared with SARS-CoV-1. *New England Journal of Medicine*, 382(16), 1564–1567.

Wang, X.-W., Li, J.-S., Jin, M., Zhen, B., Kong, Q.-X., Song, N., Xiao, W.-J.,Yin, J., Wei, W., Wang, G.-J., Si, B., Guo, B.-Z., Liu, C., Ou, G.-R., Wang, M.-N., Fang, T.-Y., Chao, F.-H., & Li, J.-W. (2005). Study on the resistance of severe acute respiratory syndrome-associated coronavirus. *Journal of Virological Methods*, 126(1–2), 171–177.

Wellings, F. M., Lewis, A. L., & Mountain, C. W. (1976). Demonstration of solids-associated virus in wastewater and sludge. *Applied and Environmental Microbiology*, 31(3), 354–358.

Wölfel, R., Corman, V. M., Guggemos, W., Seilmaier, M., Zange, S., Müller, M. A., Niemeyer, D., . . . Wendtner, C. (2020). Virological assessment of hospitalized patients with COVID-2019. *Nature*, 581, 465–469. https://doi.org/10.1038/s41586-020-2196-x

Wrapp, D., Wang, N., Corbett, K. S., Goldsmith, J. A., Hsieh, C.-L., Abiona, O., Graham, B. S., & McLellan, J. S. (2020). Cryo-EM structure of the 2019-nCoV spike in the prefusion conformation. *Science*, 367(6483), 1260–1263.

Xiao, F., Sun, J., Xu, Y., Li, F., Huang, X., Li, H., Zhao, J., Huang, H., & Zhao, J. (2020). Infectious SARS-CoV-2 in feces of patient with severe COVID-19. *Emerging Infectious Disease*, 26(8). https://doi.org/10.3201/eid2608.200681

Ye, Y., Ellenberg, R. M., Graham, K. E., & Wigginton, K. R. (2016). Survivability, partitioning, and recovery of enveloped viruses in untreated municipal wastewater. *Environmental Science & Technology*, 50, 5077–5085.

Appendix A

Example Job Safety Analysis

Job Safety Analysis

Work Package ID:

Work Package ID	JSA No.	Revision No.	2, Draft	Related Documents	SOPs and LOTO; SOC	JSA Issue Date	
Description of Work	Operation of West and East Primary Sedimentation Tanks. ** CHECK YOURSELF ** • Are you trained/certified to perform all tasks associated with this job? ARE YOU PREPARED FOR AN EMERGENCY? • Know your emergency response procedures • Verify fire extinguishers are charged & inspection tags are current • Identify emergency exits and assembly points					Expiration Date	5/19/2015
Site				Primary Treatment			
	Duty Station or Shop						
Section	Operations I Primary/Secondary	Job	Operation of West and East Primary Sedimentation Tanks				
Tools \ Safety Equipment	• Non-sparking Two Way Radio (Intrinsically safe)						
PPE Required	• Body Protection - Class III High Visibility Vest, Jacket or Equivalent • Eye Protection - Safety Glasses w/ Side Shields • Foot Protection - Work Boots / Shoes w/ Non - Slip Soles • Grippy gloves • Hand Protection - Nitrile Gloves (Disposable) • Head Protection - Hard Hat • Respiratory Protection - Dust Mask (N95 or Equivalent)			• Body Protection - Work Uniform • Eye Protection - Tight Fitting Goggles and Face Shield • Foot Protection for icy conditions - Cleated Overshoes • H2S Personal Monitor • Hand Protection - Rubber Gloves • Rain Gear			
Other Participants	None listed						
Chemicals	• Hydrogen Sulfide						

Work Activities

1. Working around Hydrogen Sulfide.

Hazards	Controls
Contact With / Inhalation of Hydrogen Sulfide	• If the Building H2S alarm is sounding or Red Light is blinking - DO NOT enter Building

JSA No: Author:

 ALLIANT

Job Safety Analysis

1. Working around Hydrogen Sulfide.

Hazards	Controls
	• Before starting work, observe the LCP to confirm ventilation is operating.
	• Ensure you wear appropriate PPE to prevent contact with / inhalation of hydrogen sulfide. **PPE Required:**
	- H2S Personal Monitor
	• If at any time during monitoring the instrument alarm sounds STOP WORK, move to a safe location and notify your Supervisor.

2. Working around Primary Sedimentation Tanks.

Hazards	Controls
Engulfment / Drowning from Falling into Tanks.	• Work within the perimeter of the handrail or guarding.
Fall / Working at Heights	• Do not lean on guardrail or fencing. If reaching over guardrails/fencing, keep your center of gravity inside guardrail perimeter.
	• When approaching edge of tank, ensure guarding is secure. Report any hazards to your Supervisor immediately.
Potential for contact with bird droppings	• Ensure you wear appropriate PPE to prevent contact with droppings. **PPE Required:**
	- Hand Protection - Nitrile Gloves (Disposable)
	- Respiratory Protection - Dust Mask (N95 or Equivalent)
Exposure to insects and spiders	• Always inspect for evidence of insects, nests and webs upon entering work area. Report infestations to supervisor immediately.
	• Avoid reaching into blind spots. If you must reach into a blind spot, wear a heavy leather glove.
	• If bitten / stung, stop work and immediately report to supervisor. Seek medical attention if needed.
Slippery surfaces caused by wet spots, spills, and ice / uneven surfaces (steps).	• Avoid walking on spills, uneven, wet and icy surfaces. Provide surface treatments as necessary (salt / sand). **PPE Required:**
	- Foot Protection - Work Boots / Shoes w/ Non - Slip Soles
	• Follow all housekeeping protocols as outlined in the Standard of Care for the Process Area.

 ALLIANT

Job Safety Analysis

3. Visually inspect the MCC's and LCP's to verify that all equipment and components are powered and ready for service.

Hazards	Controls
Release of Hazardous Energy from Electrical Panels.	• Do Not open electrical panels.
Contact with damaged or exposed electrical wires / components.	• DO NOT touch or approach an unguarded MCC, LCP or RCP enclosures which have exposed electrical components or loose wires, any open panels, broken buttons/ switches, loose screws or bolts, or open/damaged electrical connections.

If electrical panels are left open or uninsulated electrical wiring is observed, STOP! Report immediately to your Supervisor. |
| Potential for electrical shock and / or arc flash - Voltage up to 480 Volts | • ALWAYS Start and Stop Equipment using the Equipment LCP.

DO NOT operate main disconnects on the 480 Volt panel while motor is running - this will reduce the risk of arc flash. |

4. Drain the condensate from the drive units.

Hazards	Controls
Contact with Rotating Equipment - Drive Units	• Ensure Lock Out Tag Out (LOTO) procedures have been applied before performing maintenance or cleaning on energized equipment and components.
• Ensure machine guards are in place before operating equipment. STOP operating equipment with any missing or damaged guards and Report conditions to your Supervisor immediately.
• Caught / Struck by - Never place your hands or body near moving parts
• Keep loose hair secured and out of the way to avoid getting caught in machinery.
• Do not wear loose clothing and jewelry.
Ergonomics - Awkward Body Posture while accessing Drive Condensate Drain Valve.	• Ergonomics - Work at waist level, and minimize bending, reaching or twisting. Ensure you have the correct tools to complete the work activity.
• Use proper techniques and posture. Avoid stressing back and spine.

ALLIANT

Job Safety Analysis

Work Package ID:

4. Drain the condensate from the drive units.

Hazards	Controls
Ergonomics - Potential for strain/injury while exercising and/or operating valves	• Always apply force slowly to prevent injury. Sudden high force can result in injury. **PPE Required:** - Grippy gloves • Assure good footing, non-slip platforms **PPE Required:** - Foot Protection - Work Boots / Shoes w/ Non - Slip Soles • Maintain proper body position to prevent twisting or overexertion.

5. Measure the sludge blanket depth in each Sedimentation Tank using a Sludge Judge sampler at Sludge Blanket Checkpoints

Hazards	Controls
Ergonomics - Awkward Body Posture while pulling sample.	• Ergonomics - Work at waist level, and minimize bending, reaching or twisting. Ensure you have the correct tools to complete the work activity. • Use proper techniques and posture. Avoid stressing back and spine.
Exposure to pathogens from contact with sludge or wastewater.	• Ensure you wear appropriate clothing and PPE to prevent sludge from contacting skin. **PPE Required:** - Eye Protection - Tight Fitting Goggles and Face Shield - Hand Protection - Rubber Gloves

6. Cleaning or hosing down troughs, collector mechanisms, baffles, weirs, and scum hoppers with PSW.

Hazards	Controls
Splash / spray of PSW and raw wastewater.	• Ensure you wear appropriate clothing and PPE to prevent sludge / wastewater from contacting skin. **PPE Required:** - Eye Protection - Tight Fitting Goggles and Face Shield - Hand Protection - Rubber Gloves

JSA No: Author:

ALLIANT

Work Package ID:

Job Safety Analysis

6. Cleaning or hosing down troughs, collector mechanisms, baffles, weirs, and scum hoppers with PSW.

Hazards	Controls
	- Rain Gear
Potential to damage equipment or cause electrical hazards from exposure to PSW while hosing down work area.	• Do Not spray water on or near electrical receptacles or panels.

Additional Information

West Primary Sedimentation Tanks (PST-1 through PST-16) East Primary Sedimentation Tanks (PST-17 through PST-36) Operating Tasks:
• Monitor and perform daily inspections of outdoor Primary Sedimentation Tanks and Components - including Influent Sluice Gate, Influent Baffle, Center Pier and Influent Riser Pipe, Sludge and Scum Collector, Turntable and Drive Unit, Effluent Baffle, Weirs, Scum Hopper and associated piping, valves and appurtenances.
• Cleaning and light maintenance services on equipment and work areas, pick-up trash and debris, empty trash cans, wipe down hand rails and equipment, and hose down and clean equipment and work areas.

JSA No: J Author:

Job Safety Analysis

Job Safety Analysis Review Team

Printed Name	Signature	Functional Role	Concurrence Date

Job Safety Analysis Approval

Printed Name	Signature	Functional Role	Approval Date
	Signature on File	Sr. Process Engineer	2/6/2015

If required, based on Risk Score (Risk Score Total: Incomplete)

Risk score not yet completed

Task Leader

ALLIANT

Job Safety Analysis

Work Package ID:

JSA No: Author:

Job Safety Analysis Briefing

Printed Name	Signature	Functional Role	Approval Date
Assigned Workers		JSA Briefer	
DC Water Operations Personnel			

ALLIANT

Job Safety Analysis

Work Package ID:

Work Package ID	JSA No.	Revision No.	2, Draft	Related Documents	SOPs and LOTO; SOC	JSA Issue Date	

Description of Work	Operation of West and East Primary Sludge Pumping Systems - Head Houses A through I. ** CHECK YOURSELF ** • Are you trained/certified to perform all tasks associated with this job? ARE YOU PREPARED FOR AN EMERGENCY? • Know your emergency response procedures • Verify fire extinguishers are charged & inspection tags are current • Identify emergency exits and assembly points				Expiration Date	5/20/2015

	Duty Station or Shop	Primary Treatment
Site		
Section	Operations I Primary/Secondary	Job
		Operation of West and East Primary Sludge Pumping Systems Head Houses A through I
Tools \ Safety Equipment	• Flashlight	• Non-sparking Two Way Radio (Intrinsically safe)
PPE Required	• Body Protection - Class III High Visibility Vest, Jacket or Equivalent • Eye Protection - Safety Glasses w/ Side Shields • Foot Protection - Work Boots / Shoes w/ Non - Slip Soles • H2S Personal Monitor • Head Protection - Hard Hat	• Body Protection - Work Uniform • Eye Protection - Tight Fitting Goggles and Face Shield • Grippy gloves • Hand Protection - Rubber Gloves
Other Participants	None listed	
Chemicals	• Hydrogen Sulfide	

Work Activities

1. Working around Hydrogen Sulfide.

Hazards	Controls
Contact With / Inhalation of Hydrogen Sulfide	• If the Building H2S alarm is sounding or Red Light is blinking - DO NOT enter Building • Before starting work, observe the LCP to confirm ventilation is operating. Listen for running ventilation.

ALLIANT

JSA No: Author:

Job Safety Analysis

Work Package ID:

1. Working around Hydrogen Sulfide.

Hazards	Controls
	• Ensure you wear appropriate PPE to prevent contact with / inhalation of Hydrogen Sulfide. **PPE Required:** - H2S Personal Monitor • If at any time during monitoring the instrument alarm sounds STOP WORK, move to a safe location and notify your Supervisor.

2. Visually inspect the MCC's and LCP's to verify that all equipment and components are powered and ready for service.

Hazards	Controls
Release of Hazardous Energy from Electrical Panels.	• Do Not open electrical panels.
Contact with damaged or exposed electrical wires / components.	• DO NOT touch or approach an unguarded MCC, LCP or RCP enclosures which have exposed electrical components or loose wires, any open panels, broken buttons/ switches, loose screws or bolts, or open/damaged electrical connections. If electrical panels are left open or uninsulated electrical wiring is observed, STOP! Report immediately to your Supervisor.
Potential for electrical shock and / or arc flash - Voltage up to 480 Volts	• ALWAYS Start and Stop Equipment using the Equipment LCP. DO NOT operate main disconnects on the 480 Volt panel while motor is running - this will reduce the risk of arc flash.

3. Visually inspect each Primary Sludge Pump in the lower level of Head Houses A through I. Check the duplex sump pumps in the basement of each Head House.

Hazards	Controls
Flooding / Engulfment in the Basement from Water / Wastewater pipe or equipment failure.	• Always be aware of the location of the nearest exit (head house stairway). • Report any malfunctioning equipment to your Supervisor. • Always carry your Flashlight and Radio when in the Basement.

JSA No: Author:

 ALLIANT

Job Safety Analysis

Work Package ID:

3. Visually inspect each Primary Sludge Pump in the lower level of Head Houses A through I. Check the duplex sump pumps in the basement of each Head House.

Hazards	Controls
Slips, Trips and Falls	• Keep floors clean and dry. Use warning signs for wet floor areas. **PPE Required:** - Foot Protection - Work Boots / Shoes w/ Non - Slip Soles • Keep travel paths clear of obstructions and in good repair. • Follow all housekeeping protocols as outlined in the Standard of Care for the Process Area.

4. Manipulate valves to flush sludge pumps, suction and discharge piping with hard-piped PSW connections to clear any build-up of grit and heavy solids. Use caution when draining any piping that may still be pressurized with sludge and/or PSW.

Hazards	Controls

Ergonomics - Potential for strain/injury while exercising and/or operating valves	• Use a valve wrench if small handled valve cannot be gripped without bending the wrist - ONLY if space permits (potential knuckle buster).
	 PPE Required:
	    - Grippy gloves
	 • Use force reducing actuator to bring the point of motion to a safe operator position.
	 • Always apply force slowly to prevent injury. Sudden high force can result in injury.
	 • Assure good footing, non-slip platforms
	 PPE Required:
	    - Foot Protection - Work Boots / Shoes w/ Non - Slip Soles
	 • Maintain proper body position to prevent twisting or overexertion.

Release of Hazardous Energy - Sludge lines pressurized with PSW	• Ensure pressurized lines are released properly.
	 PPE Required:
	    - Eye Protection - Tight Fitting Goggles and Face Shield
	    - Hand Protection - Rubber Gloves

JSA No: Author:

ALLIANT

Job Safety Analysis

Additional Information

West Primary Sedimentation Tank Head Houses A through D - West Primary Sludge Pumping Systems (PSP-1 through PSP-16) East Primary Sedimentation Head Houses E through I - East Primary Sludge Pumping Systems (PSP-17 through PSP-36)

Operating Tasks:
• Monitor and perform daily inspections of equipment and components for leaks, worn or damaged parts and corrosion of (Piping, Pumps, Valves, Pressure Gauges and Control Panels).
• Manipulate valves and switches and observe meters and gauges. Monitor operating conditions for the West and East Primary Sludge Pumping Systems and Components.
• Cleaning and light maintenance services on equipment and work areas, pick-up trash and debris, empty trash cans, wipe down hand rails and equipment, and hose down and clean equipment and work areas.

JSA No: Author:

Job Safety Analysis

Work Package ID:

Job Safety Analysis Review Team

Printed Name	Signature	Functional Role	Concurrence Date

Job Safety Analysis Approval

Printed Name	Signature	Functional Role	Approval Date
	Signature on File	Sr. Process Engineer	2/6/2015

If required, based on Risk Score (Risk Score Total: Incomplete)

Risk score not yet completed

Task Leader

JSA No: Author:

ALLIANT

Job Safety Analysis

Work Package ID:

JSA No: Author:

Job Safety Analysis Briefing

Printed Name	Signature	Functional Role	Approval Date
Assigned Workers		JSA Briefer	
DC Water Operations Personnel			

ALLIANT

Job Safety Analysis

Work Package ID:

Work Package ID	JSA No.	Revision No.	2, Draft	Related Documents	SOPs and LOTO; SOC	JSA Issue Date	

| Description of Work | Operation of West and East Primary Scum Pumping Systems. | | | | | Expiration Date | 5/20/2015 |

** CHECK YOURSELF **
• Are you trained/certified to perform all tasks associated with this job?

ARE YOU PREPARED FOR AN EMERGENCY?
• Know your emergency response procedures
• Verify fire extinguishers are charged & inspection tags are current
• Identify emergency exits and assembly points

Site		Duty Station or Shop	Primary Treatment
Section	Operations I Primary/Secondary	Job	Operation of West and East Primary Scum Pumping Systems
Tools \ Safety Equipment	• Non-sparking Two Way Radio (Intrinsically safe)		
PPE Required	• Body Protection - Class III High Visibility Vest, Jacket or Equivalent • Eye Protection - Safety Glasses w/ Side Shields • Foot Protection - Work Boots / Shoes w/ Non - Slip Soles • Grippy gloves • Hand Protection - Nitrile Gloves (Disposable) • Head Protection - Hard Hat		• Body Protection - Work Uniform • Eye Protection - Tight Fitting Goggles and Face Shield • Foot Protection for icy conditions - Cleated Overshoes • H2S Personal Monitor • Hand Protection - Rubber Gloves • Rain Gear
Other Participants	None listed		
Chemicals	• Hydrogen Sulfide		

Work Activities

1. Working around Hydrogen Sulfide.

Hazards	Controls
Contact With / Inhalation of Hydrogen Sulfide	• If the Building H2S alarm is sounding or Red Light is blinking - DO NOT enter Building • Before starting work, observe the LCP to confirm ventilation is operating.

JSA No: Author:

ALLIANT

Job Safety Analysis

Work Package ID:

1. Working around Hydrogen Sulfide.

Hazards	Controls
	• Ensure you wear appropriate PPE to prevent contact with / inhalation of Hydrogen Sulfide. **PPE Required:** - H2S Personal Monitor • If at any time during monitoring the instrument alarm sounds STOP WORK, move to a safe location and notify your Supervisor.

2. Visually inspect the MCC's and LCP's to verify that all equipment and components are powered and ready for service.

Hazards	Controls
Release of Hazardous Energy from Electrical Panels.	• Do Not open electrical panels.
Contact with damaged or exposed electrical wires / components.	• DO NOT touch or approach an unguarded MCC, LCP or RCP enclosures which have exposed electrical components or loose wires, any open panels, broken buttons/ switches, loose screws or bolts, or open/damaged electrical connections. If electrical panels are left open or uninsulated electrical wiring is observed, STOP! Report immediately to your Supervisor.
Potential for electrical shock and / or arc flash - Voltage up to 480 Volts	• ALWAYS Start and Stop Equipment using the Equipment LCP. DO NOT operate main disconnects on the 480 Volt panel while motor is running - this will reduce the risk of arc flash.

3. Visually inspect each Primary Scum Pump and top off oil level as needed. Inspect each scum hopper for large objects and debris, and remove with a long handled hook or net. Wash down the scum hopper with PSW to dislodge a built up scum cap on the surface, and accumulation on the walls and equipment.

Hazards	Controls
Fall / Working at Heights	• Provide temporary guarding when hatch must be open. • When approaching edge of Hopper, ensure guarding is secure. Report any hazards to your Supervisor immediately. • Do not lean on guarding. If reaching over guarding, keep your center of gravity inside the

JSA No: Author:

ALLIANT

Job Safety Analysis

Work Package ID:

3. Visually inspect each Primary Scum Pump and top off oil level as needed. Inspect each scum hopper for large objects and debris, and remove with a long handled hook or net. Wash down the scum hopper with PSW to dislodge a built up scum cap on the surface, and accumulation on the walls and equipment.

Hazards	Controls
	perimeter.
Engulfment / Drowning from falling into Scum Hopper.	• Work within the perimeter of the handrail or guarding.
Splash / spray of PSW and raw wastewater.	• Ensure you wear appropriate clothing and PPE to prevent sludge / wastewater from contacting skin. **PPE Required:** - Eye Protection - Tight Fitting Goggles and Face Shield - Hand Protection - Rubber Gloves - Rain Gear
Ergonomics - Awkward Body Posture while using long handled hook or net.	• Ergonomics - Work at waist level, and minimize bending, reaching or twisting. Ensure you have the correct tools to complete the work activity. • Use proper techniques and posture. Avoid stressing back and spine.
Slippery surfaces caused by wet spots, spills, and ice / uneven surfaces (steps).	• Avoid walking on spills, uneven, wet and icy surfaces. Provide surface treatments as necessary (salt / sand). **PPE Required:** - Foot Protection - Work Boots / Shoes w/ Non - Slip Soles - Foot Protection for icy conditions - Cleated Overshoes • Follow all housekeeping protocols as outlined in the Standard of Care for the Process Area.

4. Adjust scum pump discharge pressure using pinch valves located on the intermediate level of the Head House basement.

Hazards	Controls
Ergonomics - Potential for strain/injury while exercising and/or operating valves	• Use a valve wrench if small handled valve cannot be gripped without bending the wrist - ONLY if space permits (potential knuckle buster). **PPE Required:** - Grippy gloves • Use force reducing actuator to bring the point of motion to a safe operator position. • Always apply force slowly to prevent injury. Sudden high force can result in injury.

JSA No: Author:

ALLIANT

Job Safety Analysis

Work Package ID:

4. Adjust scum pump discharge pressure using pinch valves located on the intermediate level of the Head House basement.

Hazards	Controls
	• Assure good footing, non-slip platforms **PPE Required:** - Foot Protection - Work Boots / Shoes w/ Non - Slip Soles • Maintain proper body position to prevent twisting or overexertion.
Flooding / Engulfment in the Basement from Water / Wastewater equipment failure.	• Always be aware of the location of the nearest exit (head house stairway). • Report any malfunctioning equipment to your Supervisor. • Always carry your Flashlight and Radio when working in the Basement.
Slips, Trips and Falls	• Keep travel paths clear of obstructions and traverse carefully on uneven surfaces (including damaged grating, stairs, and irregular curb heights). **PPE Required:** - Foot Protection - Work Boots / Shoes w/ Non - Slip Soles • Report leaky or faulty equipment to your Supervisor and submit Work Order to correct. • Follow all housekeeping protocols as outlined in the Standard of Care for the Process Area.

Additional Information

West Primary Sedimentation Tank Head Houses A through D - West Primary Scum Pumping Systems (PSCP-1 through PSCP-16) East Primary Sedimentation Head Houses E through I - East Primary Scum Pumping Systems (PSCP-17 through PSCP-36) Operating Tasks:
• Monitor and perform daily inspections of equipment and components for leaks, worn or damaged parts and corrosion of (Piping, Pumps, Valves, Pressure Gauges and Control Panels).
• Manipulate valves and switches and observe meters and gauges. Monitor operating conditions for the West and East Primary Scum Pumping Systems and Components.
• Cleaning and light maintenance services on equipment and work areas, pick-up trash and debris, empty trash cans, wipe down hand rails and equipment, and hose down and clean equipment and work areas.

JSA No: Author:

ALLIANT

Job Safety Analysis

Work Package ID:

Job Safety Analysis Review Team

Printed Name	Signature	Functional Role	Concurrence Date

Job Safety Analysis Approval

Printed Name	Signature	Functional Role	Approval Date
	Signature on File	Sr. Process Engineer	2/6/2015
		Task Leader	

If required, based on Risk Score (Risk Score Total: Incomplete)

Risk score not yet completed

JSA No: Author:

ALLIANT

Job Safety Analysis

Work Package ID:

Job Safety Analysis Briefing

Printed Name	Signature	Functional Role	Approval Date
Assigned Workers		JSA Briefer	
DC Water Operations Personnel			

JSA No: Author:

ALLIANT

Job Safety Analysis

Work Package ID:

Work Package ID	JSA No.	Revision No.	Related Documents	Secondary Reactor SOP; LOTO; SOC	JSA Issue Date	
		2, Draft				
Description of Work	ENR-N: OD-2, 6, 7, 12, 14, 16 - Secondary Reactors 1-6 Operation.				Expiration Date	7/16/2015
	** CHECK YOURSELF ** • Are you trained/certified to perform all tasks associated with this job? ARE YOU READY FOR AN EMERGENCY: • Know your emergency response procedures • Verify fire extinguishers are charged and inspection tags are current • Identify emergency exits and assembly points					
Site		Duty Station or Shop	Secondary Reactors			
Section	Operations I Primary/Secondary	Job	Secondary Reactor Operation - ENR-N: OD-2, 6, 7, 12, 14, 16: Reactors 1-6			
Tools \ Safety Equipment	• Non-sparking Two Way Radio (Intrinsically safe)					
PPE Required	• Body Protection - Class III High Visibility Vest, Jacket or Equivalent • Eye Protection - Safety Glasses w/ Side Shields • Foot Protection - Work Boots / Shoes w/ Non - Slip Soles • Hand Protection - Rubber Gloves		• Body Protection - Work Uniform • Eye Protection - Tight Fitting Goggles and Face Shield • Foot Protection for icy conditions - Cleated Overshoes • Head Protection - Hard Hat			
Other Participants	None listed					
Chemicals	None listed					

Work Activities

1. Working around Secondary Reactors.

Hazards	Controls
Engulfment / Drowning from falling into Reactor.	• Work within the perimeter of the handrail or guarding.

JSA No: Author:

ALLIANT

Job Safety Analysis

Work Package ID:

1. Working around Secondary Reactors.

Hazards	Controls
Fall / Working at Heights	• Do not lean on guardrail or fencing around tanks. If reaching into a tank, keep your center of gravity outside of the tank's perimeter. • When approaching edge of tank, ensure guarding is secure. Report any hazards to your Supervisor immediately.

Slippery surfaces caused by wet spots, spills, and ice / uneven surfaces (steps).	• Avoid walking on spills, uneven, wet and icy surfaces. Provide surface treatments as necessary (salt / sand).
	PPE Required:
	  - Foot Protection - Work Boots / Shoes w/ Non - Slip Soles
	  - Foot Protection for icy conditions - Cleated Overshoes
	• Follow all housekeeping protocols as outlined in the Standard of Care for the Process Area.

2. Clean Dissolved Oxygen (DO) probes: Lift probe and conduit assembly out of the reactor, clean with a PSW hose and disposable wipes, and reinstall probe and conduit assembly into reactor.

Hazards	Controls
Ergonomics - Back Injury from lifting probe	• Ergonomics - Make sure you have been instructed on proper lifting and bending techniques. Avoid twisting the body when handling material. Get assistance when necessary.

Splash / spray of PSW and raw wastewater.	• Ensure you wear appropriate clothing and PPE to prevent PSW contact with skin.
	PPE Required:
	  - Eye Protection - Tight Fitting Goggles and Face Shield
	  - Hand Protection - Rubber Gloves

Potential to damage equipment or cause electrical hazards from exposure to PSW while hosing down work area.	• Do Not spray water on or near electrical receptacles or panels.

JSA No: Author:

ALLIANT

Job Safety Analysis

Work Package ID:

Additional Information

ENR-N: OD-2, 6, 7, 12, 14, 16 Secondary Process Reactors: West 40% Flow:
• Reactor 1 (4 Pass) - West Secondary Odd - OD-14
• Reactor 2 (4 Pass) - West Secondary Even - OD-12

East 60% Flow:
• Reactor 3 (3 Pass) - East Secondary - OD-6, 7
• Reactor 4 (3 Pass) - East Secondary - OD-2, 7 • Reactor 5 (4 Pass) - East Secondary - OD-16
• Reactor 6 (4 Pass) - East Secondary - OD-16

Operating Tasks:
• Visually inspect the secondary reactors.
• Manually operate gates and valves when necessary.
• Monitor operating conditions in each reactor including Dissolved Oxygen (DO) meter readings.
• Remove floating debris from surface of reactors, pick-up trash and debris, and hose down and clean equipment and work areas.

JSA No: Author:

ALLIANT

Job Safety Analysis

Work Package ID:

Job Safety Analysis Review Team

Printed Name	Signature	Functional Role	Concurrence Date

Job Safety Analysis Approval

Printed Name	Signature	Functional Role	Approval Date
Task Leader	Signature on File	Sr. Process Engineer	2/4/2015

Signatures Based On Risk Score (Risk Score Total: 3)

JSA No: Author:

Job Safety Analysis

Work Package ID:

Job Safety Analysis Briefing

Printed Name	Signature	Functional Role	Approval Date
Assigned Workers			
DC Water Operations Personnel		JSA Briefer	

JSA No: Author:

ALLIANT

Job Safety Analysis

Work Package ID:

Work Package ID	JSA No.	Revision No.	1, Draft	Related Documents	W Secondary WAS System; LOTO; SOC	JSA Issue Date	
Description of Work	ENR-N: OD-15 West Secondary Waste Activated Sludge (WAS) System Operation Sedimentation Basins 1-12.					Expiration Date	10/2/2015
	** CHECK YOURSELF ** • Are you trained/certified to perform all tasks associated with this job? ARE YOU PREPARED FOR AN EMERGENCY? • Know your emergency response procedures • Verify fire extinguishers are charged & inspection tags are current • Identify emergency exits and assembly points						
Site				Secondary Gallery			
Section	Operations I Primary/Secondary		Duty Station or Shop	Job	West Secondary Waste Activated Sludge System Operation - ENR-N OD-10		
Tools \ Safety Equipment	• Flashlight				• Non-sparking Two Way Radio (Intrinsically safe)		
PPE Required	• Body Protection - Class III High Visibility Vest, Jacket or Equivalent • Eye Protection - Safety Glasses w/ Side Shields • Grippy gloves • Head Protection - Hard Hat				• Body Protection - Work Uniform • Foot Protection - Work Boots / Shoes w/ Non - Slip Soles • Hand Protection - Nitrile Gloves (Disposable)		
Other Participants	None listed						
Chemicals	None listed						

Work Activities

1. Traveling through and working in the Gallery.

Hazards	Controls
Flooding / Engulfment in Gallery from Water / Wastewater pipe or equipment failure.	• Always be aware of the location of the nearest exit (head house stairway). • Report any malfunctioning equipment to your Supervisor. • Always carry your Flashlight and Radio when traveling through the Galleries.

ALLIANT

JSA No: 0 Author:

Job Safety Analysis

Work Package ID:

1. Traveling through and working in the Gallery.

Hazards	Controls
Low headroom. Contact with overhead hazards such as pipes and equipment (Overhead structures and / or piping less than 7').	• Cautiously travel through work area. • Observe posted hard hat area or low headroom hazard signs. **PPE Required:** - Head Protection - Hard Hat • Report any hard hat or low headroom signs that are damaged or missing to your Supervisor for resolution.
Entanglement Hazard - From low hanging valve chain operators	• Secure low hanging valve chain operators out of the way with chain baskets and/or hooks.
Potential exposure to hazardous noise from pipes with air leaks in gallery area.	• Hearing protection required • Observe posted Ear/Hearing Protection Required hazard signs. Report any Ear/Hearing Protection Required signs that are damaged or missing to your Supervisor for resolution.
Slips, Trips and Falls	• Be aware of and avoid conduit / pipes at floor level **PPE Required:** - Foot Protection - Work Boots / Shoes w/ Non - Slip Soles • Follow all housekeeping protocols as outlined in the Standard of Care for the Process Area.

2. Visually inspect the MCC's and LCP's to verify that all equipment and components are powered and ready for service.

Hazards	Controls
Contact with damaged or exposed electrical wires / components.	• DO NOT touch or approach an unguarded MCC, LCP or RCP enclosures which have exposed electrical components or loose wires, any open panels, broken buttons/ switches, loose screws or bolts, or open/damaged electrical connections. If electrical panels are left open or uninsulated electrical wiring is observed, STOP! Report immediately to your Supervisor.
Potential for electrical shock and / or arc flash - Voltage up to 480 Volts	• ALWAYS Start and Stop Equipment using the Equipment LCP. DO NOT operate main disconnects on the 480 Volt panel while motor is running - this will reduce the risk of arc flash.
Release of Hazardous Energy from electrical panels.	• Do Not open electrical panels.

ALLIANT

JSA No: Author:

Job Safety Analysis

Work Package ID:

3. Visually inspect each Secondary RAS Pump while walking through the Secondary Gallery to verify proper operation. Configure valves as required.

Hazards	Controls
Contact with Rotating Equipment - Pump / Motor Coupling	• Ensure machine guards are in place before operating equipment. STOP operating equipment with any missing or damaged guards and Report conditions to your Supervisor immediately. • Caught / Struck by - Never place your hands or body near moving parts • Keep loose hair secured and out of the way to avoid getting caught in machinery. • Do not wear loose clothing and jewelry.

Ergonomics - Potential for strain/injury while exercising and/or operating valves	• Use a valve wrench if small handled valve cannot be gripped without bending the wrist - ONLY if space permits (potential knuckle buster).
	 PPE Required:
	   - Grippy gloves
	 • Use force reducing actuator to bring the point of motion to a safe operator position.
	 • Always apply force slowly to prevent injury. Sudden high force can result in injury.
	 • Assure good footing, non-slip platforms
	 PPE Required:
	   - Foot Protection - Work Boots / Shoes w/ Non - Slip Soles
	 • Maintain proper body position to prevent twisting or overexertion.

Additional Information

ENR-N: OD-15 West Secondary Waste Activated Sludge (WAS) System:
• Waste Sludge Pump 1 (WSP-1) - ASCC 1 - 13F in West Secondary Sedimentation Electrical Building.
• Waste Sludge Pump 2 (WSP-2) - ASCC 2 - 13F in East Secondary Sedimentation Electrical Building.
• Waste Sludge Pump 3 (WSP-3) - ASCC 3 - 1F in East Secondary Sedimentation Electrical Building.

Operating Tasks:
• Daily monitoring of the WAS pumping rate of each operating pump and the total volume of WAS sent to Dissolved Air Flotation (DAF).
• Check for adequate seal water flow to each pump once per shift.
• Check pressure gauge indications for seal water and discharge pressure for each pump.
• Adjust pump speed of each operating WAS pump to maintain the Mixed Liquor Suspended Solids (MLSS) in return sludge header to the West Secondary Reactors .

JSA No: Author:

ALLIANT

Job Safety Analysis

Work Package ID:

Job Safety Analysis Review Team

Printed Name	Signature	Functional Role	Concurrence Date

Job Safety Analysis Approval

Printed Name	Signature	Functional Role	Approval Date

If required, based on Risk Score (Risk Score Total: Incomplete)

Risk score not yet completed

Task Leader

JSA No: Author:

ALLIANT

Job Safety Analysis

Work Package ID:

JSA No: Author:

Job Safety Analysis Briefing

Printed Name	Signature	Functional Role	Approval Date
Assigned Workers		JSA Briefer	
DC Water Operations Personnel			

ALLIANT

Appendix B

Example of a Job Safety Analysis Specific to Respiratory Protection

A key industry question is whether N95 respirator, surgical or dust masks are indicated for specific job tasks. In general, surgical or dust masks are sufficient for most tasks, with rare but more hazardous exposures indicating N95 or equivalent respirators. This section provides one example of a JSA. Taken together, the JSA examples in the appendix illustrate how site- and task-specific conditions result in different PPE requirements for respiratory protection.

Job Safety Analysis	
JSA No.:	
Job/Operation Title: WWTP Primary Clarifier Entry	**Date:** March 2016
Department/Division/Section: Plant operations & Maintenance	**Analysis Developed By:**
Location(s): Wastewater Treatment plant	**Analysis Reviewed By:**
Person(s) Performing This Job: Operators & Maintenance Personnel	**Supervisor:**
Job Start Date:	**Duration:**

Task/Step	Potential Hazards	Critical Safety Practices
1. Isolate primary clarifier inlet flow	1. Electrical equipment (transformers, switching gear, breakers, high voltage lines)	1. Perform control of hazardous energy procedure on the clarifier's inlet flow. A. If controlled manually, close valve, lock and tag-out and chain valve in closed position; verify isolation. B. If inlet flow in controlled by an electronic or pneumatic control valve; perform control of hazardous energy procedure. Lock and tag-out and then remove electric and/or pneumatic power source. Once completed; close manual valve and lock and tag-out. Chain valve in the closed position. Verify isolation.
2. Drain liquid from clarifier	1. Hand tools	1. Draining process may take up to 8 hours to complete. Factor time required to drain clarifier into work schedule/plan. 2. Perform control of hazardous energy procedures on the clarifier effluent flow valves.
3. Remove man-way access panel	1. Hand tools 2. Confined space 3. Cuts	1. Once clarifier has been drained open/remove man-way access hatch. 2. Treat clarifier as a permit confined space, follow confined space entry protocols. 3. Post confined space entry permit at the entrance manway. 4. Conduct confined space atmospheric testing every time prior to entry. 5. Conduct control of hazardous energy procedure on the rake arm drive equipment.
4. Wash down/remove accumulated sludge from clarifier	1. Confined space	1. Establish a barrier around the perimeter of the sludge discharge pit. Encapsulation potential. 2. Using high pressure washer and/or fire hoses push accumulated sludge to sludge discharge pit. 3. Continue washing down the interior of the clarifier and the rake arm until they are free of any accumulated sludge.
5. Perform internal equipment inspection of rake arm and other systems as required	1. Hand tools 2. Confined space 3. Extension cords and electric powered equipment/tools	1. Perform inspection of internal systems from the clarifier floor (rake arm, scraper blades, etc.). 2. Make necessary repairs as identified.

Task/Step	Potential Hazards	Critical Safety Practices
6. Clean scum trough; inspect scum baffle and skimmer arm	1. Cranes and crane trucks 2. Hand tools 3. Ladders (portable, fixed) 4. Noise (sound pressure level), dBA	1. Cleaning scum trough is typically done with high pressure washers with employee standing in the effluent trough. NOTE: depending on the clarifier configuration; fall protection may be required (4-ft threshold). 2. Once the effluent/scum trough has been cleaned, conduct inspection and perform work as identified.
7. Conduct sandblasting and painting	1. Cranes and crane trucks 2. Hand tools 3. Ladders (portable, fixed) 4. Noise (sound pressure level), dBA 5. Confined space 6. Respiratory protection	1. Once rake arm assembly inspection and repairs have been completed, complete sandblasting and painting as required. 2. Sandblasting may require respiratory protection and additional PPE requirements (follow manufacturer recommendations/ company policy). 3. Conduct confined space atmospheric testing every time prior to entry.
8. Conduct final inspection and test run rake assemble once the repairs and painting projects have been completed	1. Painting equipment	1. Remove any and all misc. sandblasting and construction debris from clarifier. 2. Make sure scraper blades are properly adjusted. 3. Follow control of hazardous energy procedures for equipment startup. 4. Run rake arm assembly in local control to ensure proper alignment and operation. 5. Ensure all equipment has been removed from clarifier and then re-install man-way hatch.
9. Remove clarifier effluent line control of hazardous energy devices	1. Hand tools	1. Follow control of hazardous energy procedures for equipment startup and open clarifier effluent lines.
10. Open clarifier inlet flow valves	1. Hand tools	1. Remove clarifier inlet flow energy control devices in accordance with procedure. 2. Open clarifier inlet valves. 3. Start rake arm assembly. 4. Place clarifier as back in normal operation.

POTENTIAL PHYSICAL HAZARDS OF THIS JOB

Physical Hazards	Prob.	Sev.	Consequences
Cuts and lacerations to hands, body	1	2	Cuts and abrasions
Confined space	2	4	Excessive lifting, twisting, pushing, pulling, reaching, or bending
Cranes and crane trucks	2	3	Exposure to excessive light (welding)
Cuts	1	1	Falling (>4 feet)
Electrical equipment (transformers, switching gear, breakers, high voltage lines)	3	4	
Exhaust fumes	2	4	
Extension cords and electric powered equipment/tools	1	2	
Hand tools	1	2	
Inclement weather—lightning, high wind, snow, rain, sleet	1	2	
Ladders (portable, fixed)	1	2	
Noise (sound pressure level), dBA	1	2	
Painting equipment	1	2	

Severity	Probability
S-1 = Very Low	P-1 = Low
S-2 = Low	P-2 = Medium
S-3 = Medium	P-3 = High
S-4 = High	
S-5 = Very High	

HAZARD CONTROL MEASURES USED FOR THIS JOB

Administrative Controls:
CAD Printout of valve locations for shut down
Competent person/Permit authorizer
Confined space procedure
Emergency procedures
Hot work procedure
Pre-job safety brief/Inspections (pre-job)—work areas, equipment, tools, etc.
Control of hazardous energy
Monitoring (hazardous atmospheres)
Safety meetings—on-going (e.g., daily or weekly tailgate safety)

Engineering Controls:
Barrier or signage
Energy isolation device
Fall protection structures or devices
Shut/Close valve to eliminate water flow to devices being replaced/repaired
Ventilation and exhausting if needed

Required Permit(s):
Confined Entry Space Permit
Hot Work Permit
Control of Hazardous Energy

Required Training:
Confined space
Emergency plans and fire prevention plans
Energy isolation (lock-out/tag-out)
Fall Protection
Ladders, stairways, and other working surfaces
Control of Hazardous Energy
Respiratory Protection
Scaffolds (erection/inspection)
Welding, cutting, and brazing

Required PPE:
Boots—steel toe and shank, appropriate soles
Clothing—long pants, water resistant coveralls
Clothing—long sleeve shirt
Gloves—arc flash rated
Gloves—work gloves
Hard hat
Mask, N95, or other respirator; requirement varies depending on task performed and hazards that exist
Safety glasses, goggles, face shield depending on task

Other Information:
Conduct pre-job safety briefing prior to beginning clarifier shutdown.